Hallmarks on Gold
& Silver

Concise
Hallmarks on Gold
& Silver

–

William Chaffers

Wordsworth Reference

This edition published 1994 by Wordsworth Editions Ltd,
Cumberland House, Crib Street, Ware, Hertfordshire SG12 9ET.

ISBN 1-85326-348-6

Printed and bound in Denmark by Nørhaven.

The paper in this book is produced from pure wood
pulp, without the use of chlorine or any other substance
harmful to the environment. The energy used in its
production consists almost entirely of hydroelectricity
and heat generated from waste materials, thereby
conserving fossil fuels and contributing little to the
greenhouse effect.

TABLES OF DATE LETTERS
OF ASSAY OFFICES.

PREFACE.

THIS Hand-book of Hall Marks, which was first issued in 1897, has been considerably extended, and upwards of 200 new marks have been added, bringing the various alphabets up to the present time; and also giving the various other local marks. Especial care has been taken with regard to the shape of the shields enclosing the letters, and it is believed that these reproductions are accurate and reliable; and it is hoped that this little work will be useful to those requiring, in a convenient form, information respecting the marks on old gold and silver.

Our thanks are due to the Assay Masters and others who have enabled us to complete the alphabets.

Many of the letters included in the tables of London Assay Office Letters were the copyright of the late Mr. W. J. Cripps, C.B., F.S.A., author of "Old English Plate," and by the courtesy and express permission of his representatives they are used in this book.

NEW YEAR'S DAY,   C. A. M.
        1907.

England.

INTRODUCTION.

By far the most important of the English hall marks are those impressed in London. Probably out of every hundred pieces of silver plate in this country, ninety-nine were assayed at Goldsmiths' Hall. These marks are therefore first considered.

Before proceeding to notice these marks in detail, however, we have placed a chronological table showing at a glance the different times at which the marks were introduced throughout England; or such of them as were adopted in conformity with an express enactment.

Following this, we have a table showing the marks at present in use at all assay offices in the United Kingdom and Ireland. And a similar table showing the marks used in 1701-2.

Such notes on the several stamps are added, as are deemed necessary to enable those using the tables, to follow the various changes that have taken place in the number and shape of the assay marks.

Many of the letters included in the tables of London Assay Office Letters were the copyright of the late Mr. W. J. Cripps, C.B., F.S.A., author of "Old English Plate," and by the courtesy and express permission of his representatives they are used in this book.

HALL MARKS ON PLATE.

CHRONOLOGICAL TABLE OF MARKS.

1300 (28 Edw. I.). Leopard's head.

1336 (Ordinance of the Goldsmiths' Company). 1. Leopard's head crowned. 2. Owners' or goldsmiths' marks. 3. Assayer's mark, or variable letter.

1379 (2 Rich. II.). 1. Goldsmith's, "his own proper mark." 2. "Mark of the city or borough." 3. Assayer's mark, "appointed by the King."

1423 (2 Henry VI.). "Touch of the Leopard's head," and "Mark or touch of the workman."

1477 (17 Edw. IV.). "Leopard's head crowned," and "Mark of the worker." Standard of of 18-karat gold.

1573 (15 Eliz.). Standard of 22-karat gold revived.

1576 (18 Eliz.). 1. "The goldsmith to set his mark thereon." 2. "Touch of the leopard's head crowned," and "marked by the wardens."

1597 (Minutes of Goldsmiths' Company). "Her Majesty's Lion," "Alphabetical mark approved," and "Leopard's head, limited by statute."

1675 (Goldsmiths' order). "Lion," and "Leopard's head crowned, or one of them."

1697-8 (8 & 9 Will. III.). New standard of silver. "Lion's head erased; Figure of Britannia and the maker's mark, being the two first letters of his surname."

1700-1 (12 Will. III.). York, Exeter, Bristol, Chester, and Norwich appointed to assay silver plate, and stamp it with the marks of the lion's head erased and Britannia, and, in addition to the marks of their cities, a variable mark or letter in Roman character.

1701 (1 Anne). Newcastle added to the other cities for assaying and stamping plate.

1719 (6 Geo. I.). Old standard restored. The lion, leopard's head, maker's mark, and date mark, but both standards were allowed by this Act simultaneously, varying the respective marks.

1739 (12 Geo. II.). Goldsmiths' mark, "the initials of his Christian and surname."

1773 (13 Geo. III.). Birmingham and Sheffield appointed for assaying and stamping silver plate.

1784 (24 Geo. III.). Duty mark of the King's head, and drawback stamp of Britannia.

1785 (25 Geo. III.). Drawback stamp of Britannia, discontinued.

1798 (38 Geo. III.). Standard of 18-karat gold marked with a crown and 18.

1824 (5 Geo. IV.). Birmingham empowered to stamp gold.

1844 (7 & 8 Vict.). Gold of 22 karats to be stamped with a crown and 22, instead of the lion passant.

1854 (17 & 18 Vict.). Reduced standards of gold to be stamped: 15.625—12.5—9.375—for gold of 15, 12, and 9 karats, but without the crown and King's head.

1876 Foreign plate, when assayed, to be marked, in addition to the usual marks of the Hall, with the letter F in an oval escutcheon.

1890 (54 Vict.). The duty of 1s. 6d. per ounce on silver plate abolished, and the stamp of the Queen's head, duty mark, discontinued.

TABLE OF MARKS used in 1896 at the Assay Offices in England, Scotland, and Ireland.

Assay Town.	Description.	1. Quality.	2. Standard.	3. Assay Town.	4. Date.	5. Maker.
LONDON. Established 13th Century.	Gold 22 karat.	22	Crown	Leopard's head without a crown	Letter	Initials
	„ 18 „	18	Crown	Leopard's head	Letter	Initials
	„ 15 „	15.625	Nil	Leopard's head	Letter	Initials
	„ 12 „	12.5	Nil	Leopard's head	Letter	Initials
	„ 9 „	9.375	Nil	Leopard's head	Letter	Initials
	Silver O.S.	Nil	Lion passant	Leopard's head	Letter	Initials
	„ N.S.	Nil	Britannia	Lion's head erased	Letter	Initials
BIRMINGHAM. Established 1773.	Gold 22 karat.	22	Crown	Anchor	Letter	Initials
	„ 18 „	18	Crown	Anchor	Letter	Initials
	„ 15 „	15.625	Nil	Anchor	Letter	Initials
	„ 12 „	12.5	Nil	Anchor	Letter	Initials
	„ 9 „	9.375	Nil	Anchor	Letter	Initials
	Silver O.S.	Nil	Lion passant	Anchor	Letter	Initials
	„ N.S.	Nil	Britannia	Anchor	Letter	Initials
CHESTER. Re-established 1701.	Gold 22 karat.	22	Crown	Dagger and 3 sheaves	Letter	Initials
	„ 18 „	18	Crown	Dagger and 3 sheaves	Letter	Initials
	„ 15 „	15.625	Nil	Dagger and 3 sheaves	Letter	Initials
	„ 12 „	12.5	Nil	Dagger and 3 sheaves	Letter	Initials
	„ 9 „	9.375	Nil	Dagger and 3 sheaves	Letter	Initials
	Silver O.S.	Nil	Lion passant	Dagger and 3 sheaves	Letter	Initials
	„ N.S.	Nil	Britannia		Letter	Initials

	Standard	Nil / Nil	Lion passant / Britannia	Crown / Crown	Letter / Letter	Initials / Initials
SHEFFIELD. Established 1773. No gold stamped here.	Silver O.S.	Nil			Letter	Initials
	" N.S.	Nil			Letter	Initials
EDINBURGH. Established 1457.	Gold 22 karat.	22	Thistle	Castle	Letter	Initials
	" 18 "	18	Thistle	Castle	Letter	Initials
	" 15 "	15	Nil	Castle	Letter	Initials
	" 12 "	12	Nil	Castle	Letter	Initials
	" 9 "	9	Nil	Castle	Letter	Initials
	Silver O.S.	Nil	Thistle	Castle	Letter	Initials
	" N.S.	Britannia	Thistle	Castle	Letter	Initials
GLASGOW. Established 1819.	Gold 22 karat.	22	Lion rampant	Tree, fish, and bell	Letter	Initials
	" 18 "	18	Lion rampant	Tree, fish, and bell	Letter	Initials
	" 15 "	15	Lion rampant	Tree, fish, and bell	Letter	Initials
	" 12 "	12	Lion rampant	Tree, fish, and bell	Letter	Initials
	" 9 "	9	Lion rampant	Tree, fish, and bell	Letter	Initials
	Silver O.S.	Nil	Lion rampant	Tree, fish, and bell	Letter	Initials
	" N.S.	Britannia				
DUBLIN. Established 1638. No New Standard silver marked here.	Gold 22 karat.	22	Harp crowned	Hibernia	Letter	Initials
	" 20 "	20	Plume feathers	Hibernia	Letter	Initials
	" 18 "	18	Unicorn's head	Hibernia	Letter	Initials
	" 15 "	15.625	Nil	Hibernia	Letter	Initials
	" 12 "	12.5	Nil	Hibernia	Letter	Initials
	" 9 "	9.375		Hibernia	Letter	Initials
	Silver O.S.	Nil	Harp crowned			

TABLE OF MARKS *used in* 1701-2 *at the Assay Offices in England, Scotland, and Ireland.*

Assay Town	Description.	1. Quality.	2. Standard.	3. Assay Town.	4. Date.	5. Maker.
LONDON. Established 13th Century.	Gold 22 karat Silver O.S. ,, N.S.	... ...	Lion passant Lion passant Britannia	Leopard's head crowned Leopard's head crowned Lion's head erased	Letter Letter Letter	Initials Initials Initials
EXETER. Re-established 1701.	Gold 22 karat Silver O.S. ,, N.S.	Leopard's head Leopard's head Lion's head erased	Lion passant Lion passant Britannia	Castle Castle Castle	Letter Letter Letter	Initials Initials Initials
CHESTER. Re-established 1701.	Gold 22 karat Silver O.S. ,, N.S.	Leopard's head Leopard's head Lion's head erased	Lion passant Lion passant Britannia	3 demi lions and wheatsheaf ditto ditto	Letter Letter Letter	Initials Initials Initials
NEWCASTLE. Established 1702.	Gold 22 karat Silver O.S. ,, N.S.	Leopard's head Leopard's head Lion's head erased	Lion passant Lion passant Britannia	3 castles 3 castles 3 castles	Letter Letter Letter	Initials Initials Initials

YORK. Re-established 1701.	Gold 22 karat	Leopard's head	Lion passant	5 lions	Letter	Initials
	Silver O.S.	Leopard's head	Lion passant	5 lions	Letter	Initials
	,, N.S.	Lion's head erased	Britannia	5 lions on a cross	Letter	Initials
NORWICH. Re-established 1701.	Gold 22 karat	Leopard's head	Lion passant	Castle and lion	Letter	Initials
	Silver O.S.	Leopard's head	Lion passant	Castle and lion	Letter	Initials
	,, N.S.	Lion's head erased	Britannia	Castle and lion	Letter	Initials
EDINBURGH. Re-established 1631.	Gold 22 karat	Assay mark	(Thistle in 1759)	Castle	Letter	Initials
	Silver O.S.	Assay mark	...	Castle	Letter	Initials
	,, N.S.	Britannia	...	Castle	Letter	Initials
DUBLIN. Re-established 1638.	Gold 22 karat	...	Harp crowned	(Hibernia in 1730)	Letter	Initials
	Silver O.S.	...	Harp crowned	...	Letter	Initials

LONDON HALL MARKS.

THE marks on English silver stamped in London have never been more than five, and are reduced to four, although an additional mark is now placed on foreign silver assayed in England.

If we consider these marks in the order in which they were adopted, we find them in the following sequence:—

The Leopard's Head.

The Maker's Mark.

The Annual Letter.

The Lion Passant.

The Lion's Head erased, and Britannia.

The Sovereign's Head.

The Letter F.

I.—THE LEOPARD'S HEAD.

Taking first the London mark of the leopard's head, which was the earliest assay mark impressed on vessels of gold and silver, we give the forms of this stamp during the last five and a half centuries.

This mark used to be called sometimes the "Liberdes Hede," sometimes the "Liberd Heed," and sometimes the "Catte's Face." The stamp itself was known as the "punson," and it was most zealously guarded.

It is mentioned in the statute of 1300 as "une teste de leopart"; and in the charter granted in 1327 to the Goldsmiths' Company, the puncheon of the leopard's head was then said to have been of ancient use. At all events it is always found on assayed silver, from the middle of the fifteenth century.

The form of the head has changed at various times. At first the leopard's or lion's head crowned within a circle was used, and this form continued in use until early in the sixteenth century.

In 1519 the leopard's head appears with a different crown, and within a shaped outline.

From that time until the end of the seventeenth century, the crowned leopard's head was placed within a line following the shape of the head and crown. The appearance of the lion at this time is noble, and he appears as the crowned king of beasts.

In 1678 the head was once again and for the last time placed in a circle.

In 1697 the Britannia standard was introduced, and the lion's head erased was used instead of the leopard's head.

The form of this stamp has never been altered, and is still used in the same shaped outline, for the higher standard, at the present time.

The old standard was revived in 1720, and the leopard's head crowned was again used, but the shields at this time were of very uncertain shape.

In 1739 the shield was altered to a shape similar to that of the date letter.

After 1763 the head was made smaller and placed in a plain shield.

In 1822 the leopard's head was deprived of its crown, and denuded of its mane and beard—a great change from the bold front presented in the old punches; and it has ever since looked more like a half-starved cat than a lion.

Indeed, from the earliest times until 1896, this mark has been constantly changed, and each change has been for the worse.

The leopard's head of the present cycle, adopted in 1896, however, certainly is a great improvement, though the shield may not meet with universal approbation.

II.—THE MAKER'S MARK.

This mark was first made compulsory in 1363, although it was no doubt used before that time. The early workers almost invariably employed a symbol or emblem, such as an animal, fish, crown, star, or rose. It was ordered to be "a mark of the goldsmith known by the surveyor." In 1379, "Every goldsmith shall have his own proper mark upon the work." In 1433, "The mark or sign of the worker." This mark was frequently a single letter, and frequently two letters for the Christian or surname of the maker. In 1675, the Goldsmiths' order enjoins that "the plate workers shall bring their marks to Goldsmiths' Hall, and there strike the same in a table kept in the Assay Office, and likewise enter their names and places of habitation in a book there kept for that purpose, whereby the persons and their marks may be known unto the wardens of the said company." In accordance with the Act of 1697-8 the maker used the first two letters of his surname in lieu of his initials. This enactment compelled a great number of makers to obtain new punches; but in 1720, when this Act was repealed, many makers returned to their former marks. The matter was settled once and for all by the statute of 1739, which directed the makers to destroy their existing punches, and substitute the initials of their Christian and sur-names, of entirely different types from those before used.

Sometimes a small mark, such as a cross, star, etc., is found near the maker's mark; it is that of the workman, for the purpose of tracing the work to the actual maker thereof; in large manufactories some such check is indispensable.

A list of gold and silver smiths, with their names, addresses, and dates of entry at the Hall, will be found in "A HISTORY OF LONDON GOLDSMITHS AND THEIR MARKS ON PLATE, *from the earliest re-cords preserved at Goldsmiths' Hall*," by W. CHAFFERS.

III.—DATE MARK.

A letter of the alphabet. This was the assayer's mark, and was introduced in 1478, and since that time a date letter has been regularly used. The various alphabets, each composed of twenty letters, have constantly succeeded each other, different char-acters having been used at different times. At first the letters were enclosed in outlines following the shape of the letters; but since 1560 the letters have been enclosed in heraldic shields of various forms. Previous to the Restoration, the annual letter was changed on St. Dunstan's Day (19th May) when the new wardens were elected. Since 1660 the assay year commences on the 30th May, so that each letter serves for the two halves of two years. The letters J. W. X. Y. Z. are always omitted.

IV.—THE STANDARD MARK.

The standard mark of the lion passant has been used on all standard gold, and sterling silver, from

1545, until the present time, except from 1696 until 1720. The first mention of the lion passant is in the records of the Goldsmiths' Company in May, 1597, here it is called "Her Majesty's Lion." It is not referred to in any statute until 1675.

The following representations of the lion passant are of those used by the Goldsmiths' Company; the provincial marks vary slightly from those employed in London.

The lion is always represented as passant guardant, and during the first few years was life-like, crowned, and enclosed in a shaped outline.

From 1545 until 1548:—

From 1548 until 1558:—

From 1558 until 1678:—

From 1678 until 1697:—

The standard of silver was raised, and the mark of the lion passant was changed to that of "the figure of a woman commonly called Britannia," on the 25th March, 1697.

This form of stamp is still used at the present time for the higher standard.

In 1720 the old standard was again allowed and the lion passant was again used. Between 1720 and 1739 the lion was placed in a rectangle:—

From 1739 until 1756 the shaped outline was again used :—

The marks at this period are somewhat uncertain in form.

From 1756 until 1896 the lion was placed in a regular shield :—

In 1896 a new form of shield was introduced, having three lobes above and the same number below.

There are six legal standards for gold and two for silver, as follows :—

GOLD.

22 karats=917 millims.		15 karats = 625 millims.
20 ,, =834 ,, (Dublin only).		12 ,, = 500 ,.
18 ,, =750 ,,		9 ,, = 375 ,,

SILVER

11 oz. 2 dwts. = 925 millims. | 11 oz. 10 dwts.=959 millims.

FOR GOLD OF THE OLD STANDARD OF 22 KARATS, and sterling silver cf 11 oz. 2 dwts., the mark was a lion passant. Previous to 1845 there was no distinctive mark between standard gold and sterling silver. Lut in that year for gold, the lion was omitted and the quality in karats and a crown substituted.

FOR GOLD OF 18 KARATS, a crown and the figures 18, instead of the lion passant (38 Geo. III, 1798).

FOR GOLD OF 22 KARATS (or the old standard) a crown and the figures 22, instead of the lion passant (7 & 8 Vict. 1844).

FOR GOLD MANUFACTURES OF THE REDUCED STANDARD (17 & 18 Vict., 1854), the leopard's head, date letter, and the numerals.

15 karats:	15.625 on separate stamps.		
12 ,,	12.5	,,	,,
9 ,,	9.375	,,	,,

The numerals on these punches are to express, decimally, the quantity of pure gold in the article so marked, thus, pure gold being 24 karats.

15 karats $\frac{15}{24} = \frac{5}{8} = 625$ parts or millims in 1000.

12 ,, $\frac{12}{24} = \frac{4}{8} = 500$,, ,,

9 ,, $\frac{9}{24} = \frac{3}{8} = 375$,, ,,

FOR SILVER OF THE NEW STANDARD OF 11 OZ. 10 DWTS. the marks are, a figure of Britannia and the lion's head erased, instead of the lion passant and leopard's head (8 Will. III, 1697).

THE LION'S HEAD ERASED, AND FIGURE OF BRITANNIA.

When the standard for silver was raised in 1697, it was enacted that in lieu of the leopard's head and lion passant, the assay marks should be the figure of a lion's head erased, and the figure of a woman commonly called Britannia. This higher standard with these marks continued to be compulsory until 1720; when the old standard was again allowed, with the old marks. The higher standard is still perfectly legal, and when used is denoted by the lion's head erased, and figure of Britannia.

V.—THE SOVEREIGN'S HEAD.

The head in profile of the reigning sovereign. This mark indicated the payment of the duty, and was impressed at the Assay Offices on every manufactured article of standard gold and silver that was liable to the duty, after payment to the officers of the Goldsmiths' Company, who were the appointed receivers.

After the passing of the Duty Act (24 Geo. III, c. 53), which took effect on St. Dunstan's Day (30th May), 1784, the duty stamp of the *King's head in- cuse* was used for a short period. We find it in conjunction with the letter **i** of 1784, and also with the letter **k** of 1785.

The head of George III. is in an ellipse and is turned to the right :—

George the Fourth's is also turned to the right for the silver mark, though he is turned to the left on his coins :—

The next sovereign, William the Fourth, was turned to the right in a similar manner :—

The head of our late Gracious Sovereign, Queen Victoria, is turned to the left :—

The duty imposed in 1784, was—on gold, 8s. per oz.; on silver 6d. per oz. In 1797, gold, 8s. per oz.; silver, 1s. per oz. In 1804, gold, 16s. per oz.; silver, 1s. 3d. per oz. In 1815, gold, 17s. per oz.; silver, 1s. 6d. per oz.; independent of the smith's licence.

Both the *crown* and *duty mark* of the sovereign's head were omitted on the three lower standards, and although they paid the same duty as the higher standards, there was no indication of it on the stamps.

The duty on silver was abolished in 1890, and the sovereign's head consequently omitted.

VI.—FOREIGN MARK.

THE LETTER F.

In 1876 it was enacted that gold and silver plate imported from foreign parts, and assayed at any assay office in the United Kingdom, should be marked in addition to the marks used at such assay office, with the mark of the letter F on an oval escutcheon.

LONDON ASSAY OFFICE LETTERS.

LONDON ASSAY OFFICE LETTERS.

CYCLE 1. Lombardic. HENRY VII.	CYCLE 2. Black Letter Small. HENRY VIII.	CYCLE 3. Lombardic Capitals. HENRY VIII.	CYCLE 4. Roman Capitals. HENRY VIII.—MARY.
1478–9	1498–9	1518–9	1538–9
1479–0	1499–0	1519–0	1539–0
1480–1	1500–1	1520–1	1540–1
1481–2	1501–2	1521–2	1541–2
EDWARD V. 1482–3	1502–3	1522–3	1542–3
RICHARD III. 1483–4	1503–4	1523–4	1543–4
1484–5	1501–5	1524–5	1544–5
HENRY VII. 1485–6	1505–6	1525–6	1545–6
1486–7	1506–7	1526–7	1546–7
1487–8	1507–8	1527–8	EDWARD VI. 1547–8

1488-9	1508-9	1528-9	1548-9
1489-0	HENRY VIII 1509-0	1529-0	1549-0
1490-1	1510-1	1530-1	1550-1
1491-2	1511-2	1531-2	1551-2
1492-3	1512-3	1532-3	1552-3
1493-4	1513-4	1533-4	MARY. 1553-4
1494-5	1514-5	1534-5	1554-5
1495-6	1515-6	1535-6	*1555-6
1496-7	1516-7	1536-7	1556-7
1497-8	1517-8	1537-8	1557-8

THREE STAMPS.
1. Leopard's Head, crowned in 1477.
2. Date Mark.
3. Maker's Mark.
No lion passant.
No regular shield.

THREE STAMPS.
1. Leopard's Head crowned.
2. Date Mark.
3. Maker's Mark.
No lion passant.
No regular shield.

THREE STAMPS.
1. Leopard's Head crowned.
2. Date Mark.
3. Maker's Mark.
No lion passant.
No escutcheons.

FOUR STAMPS.
1. Leopard's Head crowned.
2. Date Mark.
3. Maker's Mark.
4. The lion passant first used about 1545.
No escutcheons.

* This letter being accompanied by the lion passant on plate may be distinguished from the S of 1535, when there were only three marks.

LONDON ASSAY OFFICE LETTERS.

	CYCLE 5. Black Letter Small. ELIZABETH.	CYCLE 6. Roman Capitals. ELIZABETH.	CYCLE 7. Lombardic Capitals. JAMES I.	CYCLE 8. Small Italics. CHARLES I.	CYCLE 9. Court Hand. CROMWELL.
1	ELIZABETH. 1558–9	1578–9	1598–9	1618–9	1638–9
2	1559–0	1579–0	1599–0	1619–0	1639–0
3	1560–1	1580–1	1600–1	1620–1	1640–1
4	1561–2	1581–2	1601–2	1621–2	1641–2
5	1562–3	1582–3	1602–3	1622–3	1642–3
6	1563–4	1583–4	JAMES I. 1603–4	1623–4	1643–4
7	1564–5	1584–5	1604–5	1624–5	1644–5
8	1565–6	1585–6	1605–6	CHARLES I. 1625–6	1645–6
9	1566–7	1586–7	1606–7	1626–7	1646–7
10	1567–8	1587–8	1607–8	1627–8	1647–8
11			1608–9	1628–9	1648–9

Date	Letter
1649-0	
1650-1	
1651-2	
1652-3	
1653-4	
1654-5	
1655-6	
OLIVER. 1656-7	
1657-8	

FOUR STAMPS.
1. Leopard's Head cr.
2. Lion passant.
3. Date Mark.
4. Maker's Mark.
Letter in a shield, as above.

Date	Letter
1629-0	
1630-1	
1631-2	
1632-3	
1633-4	
1634-5	
1635-6	
1636-7	
1637-8	

FOUR STAMPS.
1. Leopard's Head cr.
2. Lion passant.
3. Date Mark.
4. Maker's Mark.
Letter in a shield, as above.

Date	Letter
1609-0	
1610-1	
1611-2	
1612-3	
1613-4	
1614-5	
1615-6	
1616-7	
1617-8	

FOUR STAMPS.
1. Leopard's Head cr.
2. Lion passant.
3. Date Mark.
4. Maker's Mark.
The letter put in a shield.

Date	Letter
1588-9	
1589-0	
1590-1	
1591-2	
1592-3	
1593-4	
1594-5	
1595-6	
1596-7	
1597-8	

FOUR STAMPS.
1. Leopard's Head cr.
2. Lion passant.
3. Date Mark.
4. Maker's Mark.
The letter in a regular shield.

Date	Letter
1568-9	
1569-0	
1570-1	
1571-2	
1572-3	
1573-4	
1574-5	
1575-6	
1576-7	
1577-8	

FOUR STAMPS.
1. Leopard's Head cr.
2. Lion passant.
3. Date Mark.
4. Maker's Mark.
The date letter first put in a shield.

LONDON ASSAY OFFICE LETTERS.

CYCLE 10. Black Letter Capitals. CHARLES II.		CYCLE 11. Black Letter Small. JAS. II. & WILL. III.		CYCLE 12. Court Hand. ANNE.		CYCLE 13. Roman Capitals. GEORGE I. & II.		CYCLE 14. Roman Small. GEORGE II.	
A	*1658-9	a	1678-9	a	MAR. to MAY. 1697	A	1716-7	a	1736-7
B	1659-0	b	1679-0	b	1697-8	B	1717-8	b	1737-8
C	CHARLES II. 1660-1	c	1680-1	c	1698-9	C	1718-9	c	1738-9
D	1661-2	d	1681-2	d	1699-0	D	1719-0	d	1739-0
E	1662-3	e	1682-3	e	1700-1	E	1720-1	e	1740-1
F	1663-4	f	1683-4	f	1701-2	F	1721-2	f	1741-2
G	1664-5	g	1684-5	g	ANNE. 1702-3	G	1722-3	g	1742-3
H	1665-6	h	JAMES II. 1685-6	h	1703-4	H	1723-4	h	1743-4
I	1666-7	i	1686-7	i	1704-5	I	1724-5	i	1744-5
K	1667-8	k	1687-8	k	1705-6	K	1725-6	k	1745-6
L	1668-9	l	WILL. & MARY. 1688-9	l	1706-7	L	1726-7	l	1746-7

1669–0	1689–0	1707–8	GEORGE II. 1727–8	1747–8
1670–1	1690–1	1708–9	1728–9	1748–9
1671–2	1691–2	1709–0	1729–0	1749–0
1672–3	1692–3	1710–1	1730–1	1750–1
1673–4	1693–4	1711–2	1731–2	1751–2
1674–5	1694–5	1712–3	1732–3	1752–3
1675–6	WILLIAM III. 1695–6	1713–4	1733–4	1753–4
1676–7	30 MAY 1696	GEORGE I. 1714–5	1734–5	1754–5
1677–8	TO MAR. 1697 NO LETTER	1715–6	1735–6	1755–6

FOUR STAMPS.
1. Leopard's Head cr.
2. Lion passant.
3. Date Mark.
4. Maker's Mark.

FOUR STAMPS.
1. Leopard's Head cr.
2. Lion passant.
3. Date Mark.
4. Maker's Mark.

The leopard's head was large up to 1696; in after years it was smaller.

FOUR STAMPS.
1. Britannia.
2. Lion's Head erased.
3. Date Mark.
4. Maker's Mark.

The two first letters of the maker's surname.

FOUR STAMPS.
1. Leopard's Head cr.
2. Lion passant.
3. Date Mark.
4. Maker's Mark.

The old standard revived in 1720, but both the old and new were allowed simultaneously.

FOUR STAMPS.
1. Leopard's Head cr.
2. Lion passant.
3. Date Mark.
4. Maker's Mark.

After 1739 the initials of maker's Christian and surname.

* This letter, towards the end of the official year, appears to have been injured, as represented, but is also seen quite perfect.

NOTE.—The two stamps of the leopard's head and the lion passant were, previous to 1678, placed in irregular shields, the border line following the design; after that time the leopard's head was placed in a symmetrical shield, and the lion in a distinct oblong with a few exceptions; from and after 1750 both punches had regular heraldic shields.

LONDON ASSAY OFFICE LETTERS.

CYCLE 15. Black Letter Capitals. GEORGE III.	CYCLE 16. Roman Small. GEORGE III.	CYCLE 17. Roman Capitals. GEORGE III.	CYCLE 18. Roman Small. GEO. IV.–WILL. IV.	CYCLE 19. Black Letter Capitals. VICTORIA.
A 1756-7	a 1776-7	A 1796-7	a 1816-7	A 1836-7
B 1757-8	b 1777-8	B 1797-8	b 1817-8	B VICTORIA. 1837-8
C 1758-9	c 1778-9	C 1798-9	c 1818-9	C 1838-9
D 1759-0	d 1779-0	D 1799-0	d 1819-0	D 1839-0
E GEORGE III. 1760-1	e 1780-1	E 1800-1	e GEORGE IV. 1820-1	E 1840-1
F 1761-2	f 1781-2	F 1801-2	f 1821-2	F 1841-2
G 1762-3	g 1782-3	G 1802-3	g 1822-3	G 1842-3
H 1763-4	h 1783-4	H 1803-4	h 1823-4	H 1843-4
I 1764-5	i *1784-5	I 1804-5	i 1824-5	I 1844-5
K 1765-6	k 1785-6	K 1805-6	k 1825-6	K 1845-6
L 1766-7	l 1786-7	L 1806-7	l 1826-7	L 1846-7

Date	Mark	Date	Mark	Date	Mark	Date	Mark	Date	Mark
1767-8	𝔪	1787-8	M	1807-8	M	1827-8	m	1847-8	𝔞
1768-9	𝔫	1788-9	N	1808-9	N	1828-9	n	1848-9	𝔟
1769-0	𝔬	1789-0	O	1809-0	O	1829-0	o	1849-0	𝔠
1770-1	𝔭	1790-1	P	1810-1	P	WILL. IV. 1830-1	p	1850-1	𝔡
1771-2	𝔮	1791-2	Q	1811-2	Q	1831-2	q	1851-2	𝔢
1772-3	𝔯	1792-3	R	1812-3	R	1832-3	r	1852-3	𝔣
1773-4	𝔰	1793-4	S	1813-4	S	1833-4	s	1853-4	𝔤
1774-5	𝔱	1794-5	T	1814-5	T	1834-5	t	1854-5	𝔥
1775-6	𝔲	1795-6	U	1815-6	U	1835-6	u	1855-6	𝔦

FOUR STAMPS.
1. Leopard's Head cr.
2. Lion passant.
3. Date Mark.
4. Maker's Mark.

The leopard's head smaller after 1721 than before.

FIVE STAMPS.
1. Leopard's Head cr.
2. Lion passant.
3. Date Mark.
4. Maker's Mark.
5. King's Head.

After 1784 the duty mark of the King's head.

FIVE STAMPS.
1. Leopard's Head.
2. Lion passant.
3. Date Mark.
4. Maker's Mark.
5. King's Head.

After 1798 gold of 18 kar. was marked with a crown and 18.

FIVE STAMPS.
1. Leopard's Head.
2. Lion passant.
3. Date Mark.
4. Maker's Mark.
5. King's Head.

After 1823 the leopard's head without a crown.

FIVE STAMPS.
1. Leopard's Head.
2. Lion passant.
3. Date Mark.
4. Maker's Mark.
5. Queen's Head.

After 1845 the gold standard was marked with 22 and a crown.

* By the Duty Act of March, 1784, the payment of duty was denoted by a stamp of the King's head, which at first was *incuse* accompanied by the date letter i, and was continued in 1785-6 with the letter k; for the drawback of duty on exportation, a stamp of Britannia *incuse* was adopted, but it was discontinued in the following year; the King's head was subsequently in relief.

LONDON ASSAY OFFICE LETTERS.

CYCLE 20. Black Letter Small. VICTORIA.		CYCLE 21. Roman Capitals. VICTORIA.		CYCLE 22. Roman Small. VICTORIA.	
1856-7	a	1876-7	A	1896-7	a
1857-8	b	1877-8	B	1897-8	b
1858-9	c	1878-9	C	1898-9	c
1859-0	d	1879-0	D	1899-0	d
1860-1	e	1880-1	E	1900-1	e
1861-2	f	1881-2	F	1901-2	f
1862-3	g	1882-3	G	EDWARD VII. 1902-3	g
1863-4	h	1883-4	H	1903-4	h
1864-5	i	1884-5	I	1904-5	i
1865-6	k	1885-6	K	1905-6	k
1866-7	l	1886-7	L	1906-7	l

		1907–8
		1908–9
		1909–0

1867–8	1887–8	
1868–9	1888–9	
1869–0	1889–0	
1870–1	1890–1	
1871–2	1891–2	
1872–3	1892–3	
1873–4	1893–4	
1874–5	1894–5	
1875–6	1895–6	

FIVE STAMPS.
1. Leopard's Head.
2. Lion passant for silver.
3. Date Mark.
4. Maker's Mark.
5. Queen's Head.

For gold a crown and 22 or 18, according to standard.

FIVE STAMPS.
1. Leopard's Head.
2. Lion passant.
3. Date Mark.
4. Maker's Mark.
5. Queen's Head.

For foreign plate the letter F. Duty abolished on silver, 1890, and Queen's head omitted.

1. Leopard's Head.
2. Lion passant.
3. Date Mark.
4. Maker's Mark.

NOTE.—Large and small sized punches are used to suit the plate to be stamped; so that from 1756 to the present day, the large stamps bear the letter in a shield as here indicated—the smaller ones have the letter in a square escutcheon, the base slightly convex but not pointed, and the upper corners cut off.

England.

PROVINCIAL ASSAY OFFICES.

THE seven towns appointed by the Act 2, Henry VI. (1423) were York, Newcastle-upon-Tyne, Norwich, Lincoln, Bristol, Coventry and Salisbury, where mints had already been established, and most of them had guilds or fraternities previously existing. The town marks of the three first have been identified, but as nothing is known of the "touches" or town marks of any of the remaining four, they probably did not avail themselves of the privilege of assaying and marking plate, or if they did, no traces have been discovered of their doings or the marks they adopted.

By the Act 12 and 13 William III. (1700), York, Bristol, and Norwich and in 1701-2 Newcastle-upon-Tyne, were re-appointed, with the addition of Exeter and Chester, in which two last-named towns mints had then lately been appointed for recoining the silver moneys of the kingdom—Coventry, Salisbury, and Lincoln having then evidently ceased working.

THE HALL-MARKS OF ASSAY TOWNS.

1. LONDON. A leopard's head crowned (the ordinances of the Goldsmiths' Company of 1336, and subsequent Acts of Parliament). Since 1823 the leopard's head not crowned.

2. YORK. Five lions on a cross (discontinued).

3. EXETER. A castle with three towers (discontinued.)

4. CHESTER. Now the mark is a sword between three wheat-sheaves, but before 1779 the shield of the city arm was three demi-lions and a wheat-sheaf on a shield, and a small quartering above the sheaf.

5. NORWICH. A castle and lion passant (discontinued.)

6. NEWCASTLE. Three castles (discontinued.)

7. SHEFFIELD. A crown.

8. BIRMINGHAM. An anchor.

BARNSTABLE.

A maker, using the initials I. P., manufactured a little plate at this town in the middle of the seventeenth century.

BIRMINGHAM.

A.D. 1773. 13 GEORGE III., C. 52. This Act was passed for the appointment of Wardens and Assay Masters for assaying and stamping wrought silver plate in the towns of Sheffield and Birmingham. Silver goods "shall be marked as fclloweth; that is to say with the mark of the maker or worker thereof, which shall be the first letter of his Christian and surname; and also with the lion passant, and with the mark of the Company within whose Assay

Office such plate shall be assayed and marked, to denote the goodness thereof, and the place where the same was assayed and marked; and also with a distinct variable mark or letter, which letter or mark shall be annually changed upon the election of new wardens for each Company, to denote the year in which such plate is marked."

Sheffield and Birmingham verify their Hall-marking at the Mint, and the Act requires twice a year that the Assay Master shall appear at the Mint and verify his proceedings, under a penalty of £200, and dismissal from the office for ever, which is not the case in the other Assay Offices of Chester, Exeter, Newcastle-upon-Tyne, Edinburgh, or Dublin.

In the Parliamentary inquiry on the subject of Hall-marks and Plate in 1856, it appeared that no other offices but Birmingham and Sheffield had ever within living memory sent up their diet boxes to be tested at the Mint, being only liable when required to do so.

In the Parliamentary inquiry of 1879, it was expressly urged that the whole of the Assay Offices should be placed under the direct supervision of the Mint, so that uniform standard of quality should be guaranteed.

At Birmingham the selection of the variable letter, which is directed to be changed with the annual election of the wardens in *July*, is not confided to any officers, but the custom has been to take the letters in alphabetical order, adopting for one cycle of twenty-six years the Roman, and for another cycle the old English letters.

A.D. 1824.   5 GEORGE IV.   Power was given to the Company at Birmingham to assay gold as well as silver, and their marks are the same as London, except that the anchor is substituted for the leopard's head.

By the above-named Act of 1773, both the officers of Birmingham and Sheffield had jurisdiction to assay all plate made within twenty miles of those towns. By the 17 & 18 Victoria, cap. 96, all workers or dealers in plate are authorised to register their marks at any Assay Office legally established which they may select.

Mr. Arthur Westwood, the Assay Master at Birmingham, has most kindly furnished us with impressions of the date letters, and standard marks, now used at this city.

The following is the present form of the anchor, and of the lion passant, which is not guardant:—

BIRMINGHAM ASSAY OFFICE LETTERS.

CYCLE 1.		CYCLE 2.		CYCLE 3.		CYCLE 4.		CYCLE 5.	
Letter	JULY	Letter	JULY	Letter	JULY	Letter	JULY	Letter	JULY
A	1773-4	a	1799-0	A	1825-6	A	1850-1	a	1875-6
B	1774-5	b	1800-1	B	1826-7	B	1851-2	b	1876-7
C	1775-6	c	1801-2	C	1827-8	C	1852-3	c	1877-8
D	1776-7	d	1802-3	D	1828-9	D	1853-4	d	1878-9
E	1777-8	e	1803-4	E	1829-0	E	1854-5	e	1879-0
F	1778-9	f	1804-5	F	1830-1	F	1855-6	f	1880-1
G	1779-0	g	1805-6	G	1831-2	G	1856-7	g	1881-2
H	1780-1	h	1806-7	H	1832-3	H	1857-8	h	1882-3
I	1781-2	i	1807-8	I	1833-4	I	1858-9	i	1883-4
J	1782-3	j	1808-9	K	1834-5	K	1859-0	k	1884-5
K	1783-4	k	1809-0	L	1835-6	L	1860-1	l	1885-6
L	1784-5	l	1810-1	M	1836-7	M	1861-2	m	1886-7
M	1785-6	m	1811-2	N	1837-8	N	1862-3	n	1887-8
N	1786-7	n	1812-3	O	1838-9	O	1863-4	o	1888-9

Letter	Year	Letter	Year	Letter	Year	Letter	Year	Letter	Year
O	1787–8	o	1813–4	𝕺	1839–0	P	1864–5	𝔭	1889–0
P	1788–9	p	1814–5	𝕻	1840–1	Q	1865–6	𝔮	1890–1
Q	1789–0	q	1815–6	𝕼	1841–2	R	1866–7	𝔯	1891–2
R	1790–1	r	1816–7	𝕽	1842–3	S	1867–8	𝔰	1892–3
S	1791–2	s	1817–8	𝕾	1843–4	T	1868–9	𝔱	1893–4
T	1792–3	t	1818–9	𝕿	1844–5	U	1869–0	𝔲	1894–5
U	1793–4	u	1819–0	𝖀	1845–6	V	1870–1	𝔴	1895–6
V	1794–5	v	1820–1	𝖁	1846–7	W	1871–2	𝔪	1896–7
W	1795–6	w	1821–2	𝖂	1847–8	X	1872–3	𝔵	1897–8
X	1796–7	x	1822–3	𝖃	1848–9	Y	1873–4	𝔶	1898–9
Y	1797–8	y	1823–4	𝖅	1849–0	Z	1874–5	𝔷	1899–0
Z	1798–9	z	1824–5						

FIVE STAMPS.
1. Anchor.
2. Lion passant.
3. Date Mark.
4. Sovereign's Head, from 1784.
5. Maker's Initials.

FIVE STAMPS.
1. Anchor.
2. Lion passant.
3. Date Mark.
4. Sovereign's Head.
5. Maker's Initials.

FIVE STAMPS.
1. Anchor.
2. Lion passant.
3. Date Mark.
4. Sovereign's Head.
5. Maker's Initials.

FIVE STAMPS.
1. Anchor.
2. Lion passant.
3. Date Mark.
4. Queen's Head.
5. Maker's Initials.

FIVE STAMPS.
1. Anchor.
2. Lion passant.
3. Date Letter.
4. Queen's Head.
5. Maker's Initials.

Duty abolished on silver 1890, and Queen's head omitted.

NOTE.—For the New Standard of 11 oz. 10 dwts. a stamp of Britannia is used instead of the Lion passant.

BIRMINGHAM ASSAY OFFICE LETTERS.

CYCLE 6.

a 1900–1	**c** 1902–3	**e** 1904–5	**g** 1906–7	**i** 1908–9	
b 1901–2	**d** 1903–4	**f** 1905–6	**h** 1907–8	**k** 1909–0	

1. Anchor.
2. Lion passant.
3. Date Letter.
4. Maker's Mark.

BRISTOL.

Bristol may, perhaps, have had an office, for there were several silversmiths there who afterwards sent their goods to Exeter to be assayed.

It is not, however, by any means certain that the right of assay was ever exercised at Bristol; although it was appointed as an assay town in 1423, and re-appointed in 1700. Indeed though we have en-quired from a leading silversmith at Bristol, we have failed to trace any local silver.

There is a cup on a stem, ornamented with punched diamond pattern, which, from the inscription, ap-pears to have been made in this town, although it bears no Hall-mark. It is late sixteenth century work : —

In the possession of
Sir A. H. Elton, Bart.

(" From Mendep I was brought,
Out of a leden mine ;
In Bristol I was wrought,
And now am silver fine."

There are some interesting pieces of plate pre-served by the Corporation of Bristol, especially a pair of gilt tankards, richly decorated, the gift of John Dodridge, Recorder of Bristol, 1658, and a gilt ewer and salver, the gift of Robert Kitchen. These were both assayed and marked in London. The salver made in 1595 was stolen during the Bristol riots in October, 1831, and was cut up into 167 pieces, in which state it was offered for sale to a silversmith of the town, who apprehended the thief, and he was sentenced to fourteen years' transporta-tion. The pieces were rivetted together on a silver plate by the same silversmith, in which state it now remains, its history being recorded on the back. A State sword, bearing date 1483, ornamented and enamelled, is also preserved ; on one of the mounts are the arms of Bristol, viz., a three-masted ship

approaching a castle on a rock, with two unicorns
as supporters, and on a torse two arms, one holding
a serpent, the other the scales of justice, being the
crest of the city arms.

CHESTER.

It appears by the record of Domesday, that in the
reign of Edward the Confessor there were seven
Mint Masters in Chester. In the reign of Charles I.
much of the silver was coined here, and in that of
William III. it was one of the six cities in which
mints were established for recoining the silver of
the kingdom. The Mint-mark of Chester on the
half-crowns of Charles I. struck in 1645 is three
gerbes or wheat-sheaves.

We have no record of the time when Chester first
commenced assaying plate; it is not mentioned in
the statute of the 2nd Henry IV. (A.D. 1423), but an
office must have been established early in the sixteenth century. An old minute-book contains an
entry some time prior to 1573, directing "that noe
brother shall delevre noe plate by him wrought unles
his touche be marked and set upon the same before
deliverie thereof, upon paine of forfeiture of everie
diffalt to be levied out of his goods iij^s iiij^d."

The arms of Chester, granted in 1580, were, party
per pale, composed of the dexter half of the coat of
England, *gules*, three lions passant guardant dimidiated, *or*, and the sinister half of the coat of Blunde-
ville, Earl of Chester, *azure*, three gerbes also dimidiated *or*. The crest is, on a wreath *or gules* and
azure, over a royal helmet, a sword of State erect,
with the point upwards. Supporters: on the dexter
side, a lion rampant *or*, ducally gorged *argent;* on
the sinister side, a wolf *argent*, ducally gorged *or*.
The grant mentions the antiquity of the city, &c., and

that the ancient arms were nearly lost by time and negligence, and that the coat which the citizens claimed was deficient in crest and supporters. The Hall marks on plate were the arms of the city, a dagger erect between three wheat-sheaves, down to 1697. In 1701, the shield adopted was three demi-lions with three wheatsheaves also dimidiated, which was again changed about 1775 to the more simple shield above described, without the demi-lions, &c., still in use.

Chester was re-appointed by the Act 12th William III. (1700), and is regulated by that Act and the statute of 12th George II.

The variable letter was changed annually on the 5th July, from 1701 until 1839; it was then changed on the 5th August until 1890; since which time the change has been made annually on the 1st July.

Chester has, since 1889, voluntarily submitted its Diet for assay at the Mint, at the same time as the Birmingham and Sheffield Diets are verified.

The following is the present form of the Chester Mark :—

CHESTER ASSAY OFFICE LETTERS.

CYCLE 1.		CYCLE 2.		CYCLE 3. JULY		CYCLE 4. JULY		CYCLE 5. JULY	
1664–5	A	1689–0	A	1701–2	A	1726–7	A	1752–3	A
1665–6	B	1690–1	B	1702–3	B	1727–8	B	1753–4	B
1666–7	C	1691–2	C	1703–4	C	1728–9	C	1754–5	C
1667–8	D	1692–3	D	1704–5	D	1729–0	D	1755–6	D
1668–9	E	1693–4	E	1705–6	E	1730–1	E	1756–7	E
1669–0	F	1694–5	F	1706–7	F	1731–2	F	1757–8	F
1670–1	G	1695–6	G	1707–8	G	1732–3	G	1758–9	G
1671–2	H	1696–7	H	1708–9	H	1733–4	H	1759–0	H
1672–3	I			1709–0	I	1734–5	I	1760–1	I
1673–4	K			1710–1	K	1735–6	J	1761–2	J
1674–5	L			1711–2	L	1736–7	K	1762–3	K
1675–6	M			1712–3	M	1737–8	L	1763–4	L
1676–7	N			1713–4	N	1738–9	M	1764–5	M
1677–8	O			1714–5	O	1739–0	N	1765–6	N

Letter	Year
ℙ	1678-9
ℚ	1679-0
ℝ	1680-1
𝕊	1681-2
𝕋	1682-3
𝕌	1683-4
𝕍	1684-5
𝕎	1685-6
𝕏	1686-7
𝕐	1687-8
ℤ	1688-9

FOUR MARKS.
1. City Arms, of a dagger between 3 gerbes.
2. Crest, a sword erect.
3. Date Letter.
4. Maker's Mark.

In a minute of 1686 three Hall-marks are mentioned, that of the Maker making four.

From 1697 to 1701 the New Standard was only stamped in London; the Old Standard being illegal, the Provincial Offices could not assay or stamp plate.

Letter	Year
P	1715-6
Q	1716-7
R	1717-8
S	1718-9
T	1719-0
U	1720-1
V	1721-2
W	1722-3
X	1723-4
Y	1724-5
Z	1725-6

FOUR MARKS.
1. City Arms, as before.
2. Crest, fleur-de-lis, or sword erect.
3. Date Letter.
4. Maker's Mark.

FIVE MARKS.
1. City Arms, changed about 1720 to 3 demi-lions and 3 half gerbes.
2. Britannia.
3. Leopard's Head cr.
4. Date Letter.
5. Maker's Mark.
After 1720, Old Standard.

Letter	Year
O	1740-1
P	1741-2
Q	1742-3
R	1743-4
S	1744-5
T	1745-6
U	1746-7
V	1747-8
W	1748-9
X	1749-0
Y	1750-1
Z	1751-2

FIVE MARKS.
1. City Arms, as the preceding, after 1720.
2. Lion passant.
3. Leopard's Head.
4. Date Letter.
5. Maker's Mark.

Letter	Year
O	1766-7
*P	1767-8
Q	1768-9
R	1769-0
S	1770-1
T	1771-2
U	1772-8
W	1773-4
X	1774-5
Y	1775-6
Z	1776-7

FIVE MARKS.
1. Lion passant.
2. Leopard's Head.
3. City Arms, as the preceding.
4. Date Letter.
5. Maker, as before.

NOTE.—The letters after 1701, with few exceptions, are placed in square escutcheons, with the corners cut off.

* Sir Philip Egerton, of Oulton, has sent us lac-similes of the Chester Marks on a pair of barrel-mugs, with P in Roman capitals, and an invoice of R. Richardson, Silversmith, 1769, made in 1767-8 for P. Egerton, Esq., of Oulton.

CHESTER ASSAY OFFICE LETTERS.

Letter	CYCLE 6	CYCLE 7	CYCLE 8	CYCLE 9	CYCLE 10
a / A	1777–8	JULY 1797–8	JULY 1818–9	JULY 1839–0	JULY 1864–5
b / B	1778–9	1798–9	1819–0	1840–1	1865–6
c / C	1779–0	1799–0	1820–1	1841–2	1866–7
d / D	1780–1	1800–1	1821–2	1842–3	1867–8
e / E	1781–2	1801–2	1822–3	1843–4	1868–9
f / F	1782–3	1802–3	1823–4	1844–5	1869–0
g / G	1783–4	1803–4	1824–5	1845–6	1870–1
h / H	1784–5	1804–5	1825–6	1846–7	1871–2
i / I	1785–6	1805–6	1826–7	1847–8	1872–3
k / K	1786–7	1806–7	1827–8	1848–9	1873–4
l / L	1787–8	1807–8	1828–9	1849–0	1874–5
m / M	1788–9	1808–9	1829–0	1850–1	1875–6
n / N	1789–C	1809–0	1830–1	1851–2	1876–7
o / O	1790–1	1810–1	1831–2	1852–3	1877–8

1791-2	1811-2	1832-3	1853-4	1878-9
1792-3	1812-3	1833-4	1854-5	1879-0
1793-4	1813-4	1834-5	1855-6	1880-1
1794-5	1814-5	1835-6	1856-7	1881-2
1795-6	1815-6	1836-7	1857-8	1882-3
1796-7	1816-7	1837-8	1858-9	1883-4
	1817-8	1838-9	1859-0	
			1860-1	
			1861-2	
			1862-3	
			1863-4	

p	$\mathscr{P}$	P	𝕬	𝕻
q	$\mathscr{Q}$	Q	𝕺	𝕼
r	$\mathscr{R}$	R	𝕽	𝕽
s	$\mathscr{S}$	S	𝕾	𝕾
t	$\mathscr{T}$	T	𝕿	𝕿
u	$\mathscr{U}$	U	𝖀	𝖀
	$\mathscr{V}$	V	𝖁	

The Stamp of the City Arms of 3 demi-lions and gerbe, changed to the Old Stamp of a sword between three gerbes, about 1784.

Six Marks.
1. Lion passant.
2. Leopard's Head.
3. City Arms.
4. Date Letter.
5. Duty Mark in 1784.
6. Maker's Mark.
These letters are not facsimiles.

Six Marks.
1. Lion passant.
2. Leopard's Head.
3. City Arms.
4. Duty Mark.
5. Date Mark.
6. Maker.

Six Marks.
1. Lion passant.
2. Leopard's Head.
3. City Arms.
4. Duty Mark.
5. Date Mark.
6. Maker.

Five Marks.
1. Lion passant.
2. City Arms.
3. Duty Mark.
4. Date Mark.
5. Maker.
(The Leopard's Head discontinued 1839).

Five Marks.
1. Lion passant.
2. City Arms.
3. Duty Mark.
4. Date Mark.
5. Maker.

CHESTER ASSAY OFFICE LETTERS.

CYCLE 11.

Letter	Date	Letter	Date
A	1884–5	K	1893–4
B	1885–6	L	1894–5
C	1886–7	M	1895–6
D	1887–8	N	1896–7
E	1888–9	O	1897–8
F	1889–0	P	1898–9
G	1890–1	Q	1899–0
H	1891–2	R	1900–1
I	1892–3		

CYCLE 12.

Letter	Date	Letter	Date
A	1901–2	E	1905–6
B	1902–3	F	1906–7
C	1903–4	G	1907–8
D	1904–5	H	1908–9

1. Lion passant.
2. City Arms.
3. Date Mark.
4. Maker's Mark.
5. Queen's Head.

Duty abolished on silver in 1890, and Queen's Head omitted.

EXAMPLES.

CHESTER, 1665. The following four marks occur on a porringer or two-handled cup and cover, lately in the possession of *Messrs. Lewis and Son*, Brighton. It is the earliest authentic piece of Chester plate we have hitherto met with, enabling us to ascertain the type of letter used in the cycle commencing 1664.

1. The Chester City Arms, a sword between three wheatsheaves or gerbes.

2. The City Crest, adopted by the Assay Office as their Hallmark formerly, viz., a sword with a bandelet, which is still used by the officials on their printed documents, issuing from an earl's coronet, the five pellets underneath indicating the balls of the coronet.

City Crest.
Still used by the Assay Office as a heading to letters and correspondence.

3. A German text B, denoting the year 1665.

4. The maker's initials crowned, probably some of the Pembertons, who were silversmiths at Chester and members of the guild about that date.

CHESTER, 1689. These marks are on a spoon with flat stem, leaf-shaped end, rat-tail bowl, clearly of this date. In the possession of the *Earl of Breadalbane.*

1. The Chester City Arms of a sword between three gerbes or wheat-sheaves.
2. The Crest of the Assay Office at Chester.
3. Court-hand *A*, denoting the year 1689, according to the minutes of the year 1690.
4. The maker's initials, Alexander Pulford, silver-smith, who was admitted in that year as a member of the guild, whose name occurs frequently in the minutes.

The assay mark of a fleur-de-lis, somewhat similar to the sword and bandelet, requires some explanation; and Mr. Lowe, the Assay Master, remarks as a strange coincidence, that in the same old minute-book there is a sketch of a fleur-de-lis, as above shown, from which we may infer that this stamp was an old Chester mark, and we may with some degree of certainty attribute the stamp of a fleur-de-lis within a circle, so frequently found on plate of the early part of the seventeenth century, to Chester, when some such distinctive mark must have been used, and the lis has never hitherto been accounted for.

UNCERTAIN CHESTER MARKS.

CIRCA, 1660. A rat-tail spoon in the possesion of the *Rev. T. Staniforth.*
A piece of plate of the seventeenth century in *Messrs. Hancock's* possession.

COVENTRY.

Although this city is mentioned in the statute of 2nd Henry VI. as being entitled to assay plate, it is not probable that plate was ever assayed here.

EXETER.

There are no records at this Hall previous to 1701. The early mark used at Exeter before this date, was a letter X crowned, subsequently altered to a castle of three towers. The Act passed in 1700, reappointing this city for assaying plate, did not come into operation until the 29th September, 1701. On the 7th of August the Company of Goldsmiths met, and on the 17th of September Wardens were appointed, and they resolved, with all convenient speed and safety to put the Act in execution; and the first assayer was sworn in before the Mayor on the 19th of November, 1701. The letters commenced with a Roman capital A for that year, as ordered by the statute, which characters, large and small, they used throughout the alphabet until 1837, when they adopted old English capitals for that cycle. A Table of Letters for each year will be found annexed.

At this Office only one standard of gold was assayed, which was the highest standard of 22 karats.

Since 1701 the date letter has always been changed on the 7th of August in each year.

The office at this city continued to do useful work, until twenty years ago, when it was closed. A great part of the silver assayed at Exeter was manufactured in Bristol.

Ultimately the amount of business decreased to so large an extent that on the 26th June, 1883, a special Court was held at the Goldsmiths' Hall. At this Court there were present Mr. Josiah Williams, Mr.

John Ellett Lake, Mr. Ross, Mr. Henry Lake, Mr. Maynard, Assay Master, and Mr. Henry Wilcocks Hooper, Solicitor to the Company. The Company resolved, having regard to the small quantity of silver recently marked, that it was not desirable to obtain new punches; and that the premises used for the business should be given up; and that no fresh premises should be taken until sufficient applications were received to render it desirable to re-open the Hall The old punches were surrendered to the Inland Revenue Office, and the books and papers deposited with Mr. Hooper, the Solicitor to the Company.

The early minute books and other documents of the Company are now in the custody of Mr. Hooper; and six copper plates, on which many of the date letters and makers' marks have been struck, are now in the custody of Mr. J. Jerman, of Exeter.

For much of this information relating to the Exeter Assay Office, we are indebted to Mr. Percy H. Hooper, the last Deputy Assayer, and Mr. J. Jerman.

The form of the castle used at Exeter has varied at different times. At first the mark appeared of the following form: —

About 1710 the form was slightly varied: —

In 1823 the three towers are detached and placed in an oblong : —

A few years later the castles were again joined, and that form was retained until the office was closed : —

The lion passant was very similar to that used at Birmingham : —

The arms of the City of Exeter are : —Per pale *gu* and *sa* a triangular castle with three towers *or*. *Crest* a demi-lion rampant *gu*, crowned, *or*, holding between its paws a bezant, surmounted by a cross botonne *or*. Supporters, two pegasi *ar*, wings endorsed, maned, and crined *or;* on the wings three bars wavy *az*. Motto, "Semper Fidelis."

EXETER ASSAY OFFICE LETTERS.

	CYCLE 1.		CYCLE 2.		CYCLE 3.		CYCLE 4.
A	AUGUST 1701-2	a	AUGUST 1725-6	A	AUGUST 1749-0	a	AUGUST 1773-4
B	1702-3	b	1726-7	B	1750-1	b	1774-5
C	1703-4	c	1727-8	C	1751-2	c	1775-6
D	1704-5	d	1728-9	D	1752-3	d	1776-7
E	1705-6	e	1729-0	E	1753-4	e	1777-8
F	1706-7	f	1730-1	F	1754-5	f	1778-9
G	1707-8	g	1731-2	G	1755-6	g	1779-0
H	1708-9	h	1732-3	H	1756-7	h	1780-1
I	1709-0	i	1733-4	I	1757-8	i	1781-2
K	1710-1	k	1734-5	K	1758-9	k	1782-3
L	1711-2	l	1735-6	L	1759-0	l	1783-4
M	1712-3	m	1736-7	M	1760-1	m	1784-5

Letter	Year	Letter	Year	Letter	Year	Letter	Year
N	1713-4	n	1737-8	N	1761-2	n	1785-6
O	1714-5	o	1738-9	O	1762-3	o	1786-7
P	1715-6	p	1739-0	P	1763-4	p	1787-8
Q	1716-7	q	1740-1	Q	1764-5	q	1788-9
R	1717-8	r	1741-2	R	1765-6	r	1789-0
S	1718-9	f	1742-3	S	1766-7	f	1790-1
T	1719-0	t	1743-4	T	1767-8	t	1791-2
V	1720-1	v	1744-5	V	1768-9	v	1792-3
W	1721-2	w	1745-6	W	1769-0	w	1793-4
X	1722-3	x	1746-7	X	1770-1	x	1794-5
Y	1723-4	y	1747-8	Y	1771-2	y	1795-6
Z	1724-5	z	1748-9	Z	1772-3	z	1796-7

Five Stamps.
1. Lion's Head erased.
2. Britannia.
3. Castle.
4. Date Mark.
5. Maker's Initials.
[In 1720 the marks of

Five Stamps.
1. Lion passant.
2. Leopard's Head.
3. Castle.
4. Date Mark.
5. Maker's Initials.
old Standard resumed.]

Five Stamps.
1. Lion passant.
2. Leopard's Head.
3. Castle.
4. Date Mark.
5. Maker's Initials.

Five Stamps.
1. Lion passant.
2. Castle.
3. Date Mark.
4. Maker's Initials.
5. Duty Mark of King's Head in 1784.

EXETER ASSAY OFFICE LETTERS.

CYCLE 5.		CYCLE 6.		CYCLE 7.		CYCLE 8.	
AUGUST 1797–8	A	AUGUST 1817–8	a	AUGUST 1837–8	A	AUGUST 1857–8	A
1798–9	B	1818–9	b	1838–9	B	1858–9	B
1799–0	C	1819–0	c	1839–0	C	1859–0	C
1800–1	D	1820–1	d	1840–1	D	1860–1	D
1801–2	E	1821–2	e	1841–2	E	1861–2	E
1802–3	F	1822–3	f	1842–3	F	1862–3	F
1803–4	G	1823–4	g	1843–4	G	1863–4	G
1804–5	H	1824–5	h	1844–5	H	1864–5	H
1805–6	I	1825–6	i	1845–6	I	1865–6	I
1806–7	K	1826–7	k	1846–7	K	1866–7	K
1807–8	L	1827–8	l	1847–8	L	1867–8	L

Letter	Year	Letter	Year	Letter	Year	Letter	Year
M	1808–9	m	1828–9	M	1848–9	M	1868–9
N	1809–0	n	1829–0	N	1849–0	N	1869–0
O	1810–1	o	1830–1	O	1850–1	O	1870–1
P	1811–2	p	1831–2	P	1851–2	P	1871–2
Q	1812–3	q	1832–3	Q	1852–3	Q	1872–3
R	1813–4	r	1833–4	R	1853–4	R	1873–4
S	1814–5	s	1834–5	S	1854–5	S	1874–5
T	1815–6	t	1835–6	T	1855–6	T	1875–6
U	1816–7	u	1836–7	U	1856–7	U	1876–7

FIVE STAMPS.
1. Lion passant.
2. Castle.
3. King's Head.
4. Date Letter.
5. Maker's Initials.

FIVE STAMPS.
1. Lion passant.
2. Castle.
3. King's Head.
4. Date Letter.
5. Maker's Initials.

FIVE STAMPS.
1. Lion passant.
2. Castle.
3. Queen's Head.
4. Date Letter.
5. Maker's Initials.

FIVE STAMPS.
1. Lion passant.
2. Castle.
3. Queen's Head.
4. Date Letter.
5. Maker's Initials.

EXETER ASSAY OFFICE LETTERS.

CYCLE 9.

1877-8	1878-9	1879-0	1880-1	1881-2
A	**B**	**C**	**D**	**E**

1882-3
F

1. Lion passant.
2. Castle.
3. Queen's Head.
4. Date Letter.
5. Maker's Initials.

EXAMPLES.

 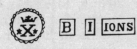 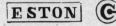

Apostle spoon, date about 1576.—*Messrs. Hancock.*

A spoon of the sixteenth century, with hexagonal stem, pear-shaped bowl, button top. Date of presentation 1620.— *Earl of Breadalbane.*

Apostle spoon, 1637.— *Rev. T. Staniforth.*

A spoon of about 1670, flat stem and oval bowl, bears this stamp with monogram and maker's initials W. F.—*Earl of Breadalbane.*

Split head spoon, pricked $^{EP}_{MN}$ 1689. *Circâ* 1689.—*Messrs. Ellett Lake & Son.*

Handsome tankard. Date 1703.—*Messrs. Ellett Lake & Son.*

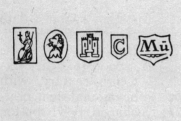

Date 1703. These new standard marks are on a three-pint tankard. (Britannia holds in her hand a flower or sprig, not a cross as here given in the cut.)—*Messrs. Hancock.*

Salver *circâ* 1710. The City mark of a Castle has a thin line rising from the pointed base of the shield to the central tower, indicating the partition *per pale*, like the City arms.— *Messrs. Ellett Lake & Son.*

Split head spoon. Date 1711.—*Messrs. Ellett Lake & Son.*

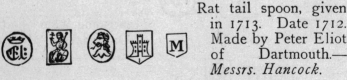

Rat tail spoon, given in 1713. Date 1712. Made by Peter Eliot of Dartmouth.— *Messrs. Hancock.*

HULL.

A little plate was marked here with the town arms during the seventeenth century, though there was never a proper Assay Office at this place.

EXAMPLE.

Spoon. Date *circâ* 1660.— *J. H. Walter, Esq.*

LINCOLN.

This city was mentioned as an assay town in 1423, but it does not appear that plate was ever hall marked here, or indeed manufactured to any large amount.

NEWCASTLE-UPON-TYNE.

At Newcastle-upon-Tyne as early as 1249, Henry III. commanded the bailiffs and good men to choose four of the most prudent and trusty men of their town for the office of moneyers there, and other four like persons for keeping the King's Mint in that town; also, two fit and prudent goldsmiths to be assayers of the money to be made there.

By the Act of 1423 this town was appointed one of the seven provincial assay towns in England.

In 1536 the goldsmiths were, by an ordinary, incorporated with the plumbers and glaziers, and the united company required to go together, on the feast of Corpus Christi, and maintain their play of the "'Three Kings of Coleyn." They were to have four wardens, viz., one goldsmith, one plumber, one glazier, and one pewterer or painter; and they had their hall in "Maden Tower" granted them in the mayoralty of Sir Peter Riddell in 1619, and the association of the goldsmiths with the other tradesmen seems to have lasted till 1702.*

This town was reappointed as an assay town by the Act of 1701.

The annual letter appears to have been used from 1702. Mr. Thomas Sewell, one of the Wardens of the Assay Office, has kindly furnished us with a Table of Date-Letters, chronologically arranged, compiled from the Assay Office Books and the copperplate on which the maker strikes his initials, as well as from pieces of old plate which have from time to time come under his notice. From careful examination of various examples of Newcastle plate, we have, in this edition, altered some of the characters, making the table more complete. The change of letter took place on the 3rd of May in each year.

In 1773 the members of the Goldsmiths' Company at Newcastle-upon-Tyne were—Mr. John Langlands and Mr. John Kirkup, Goldsmiths and Silversmiths, Wardens; and Mr. Matthew Prior, Assayer.

The names and places of abode of all the Goldsmiths, Silversmiths, and Plate-workers then living, who had entered their names and marks were—Mr. John Langlands, Mr. John Kirkup, Mr. Samuel James, Mr. James Crawford, Mr. John Jobson, Mr. James Hetherington (Newcastle-upon-Tyne), Mr. John Fearney (Sunderland), and Mr. Samuel Thomson (Durham).

The Assay Office at Newcastle was closed in May, 1884, in consequence of there being insufficient work to make it worth keeping open. The Assay Master of the Office before 1854 was Mr. F. Somerville. He

---

* From an Impartial History of the Town and County of Newcastle-upon-Tyne, published in 1801, p. 429.

was succeeded by Mr. James Robson, who entered the office as a stamper in 1836, became Assay Master in 1854, and retained that post until the Office was finally closed. The last two wardens were Mr. T. A. Reid and Mr. J. W. Wakinshaw. A curious incident occurred when Mr. Robson commenced his duties. By some means he obtained the wrong punches, and marked some plate which afterwards went to Carlisle. This almost led to an action against a silversmith at that city, who was accused of forging the hall marks.

When the office was closed the stamping punches were obliterated or defaced by an Inland Revenue Officer. The name punch plate and the old books of the Goldsmiths' Company were placed in the Black Gate Museum of the Old Castle in the city.

NEWCASTLE-UPON-TYNE ASSAY OFFICE LETTERS.

CYCLE 1. MAY		CYCLE 2. MAY		CYCLE 3. MAY		CYCLE 4. MAY	
A	1702–3	A	1724–5	A	1746–7	A	1769–0
B	1703–4	B	1725–6	B	1747–8	B	1770–1
C	1704–5	C	1726–7	C	1748–9	C	1771–2
D	1705–6	D	1727–8	D	1749–0	D	1772–3†
E	1706–7	E	1728–9	E	1750–1	E	1773–4
F	1707–8	F	1729–0	F	1751–2	F	1774–5
G	1708–9	G	1730–1	G	1752–3	G	1775–6
H	1709–0	H	1731–2	H	1753–4	H	1776–7
I	1710–1	I	1732–3	I	1754–5	I	1777–8
K	1711–2	K	1733–4	J	1755–6	K	1778–9
				K	1756–7	C	1779–0

Letter	Year	Letter	Year	Letter	Year	Letter	Year
L	1712–3	L	1734–5	L	1757–8		
M	1713–4	M	1735–6	M	1758–9	M	1780–1
N	1714–5	N	1736–7	N	1759–0	N	1781–2
O	1715–6	O	1737–8	O	1760–1	O	1782–3
P	1716–7	P	1738–9	P	1761–2	P	1783–4
Q	1717–8	Q	1739–0	Q	1762–3	Q	1784–5
R	1718–9	R	1740–1	R	1763–4	R	1785–6
S	1719–0	S	1741–2	S	1764–5	S	1786–7
T	1720–1	T	1742–3	T	1765–6	T	1787–8
U	1721–2	U	1743–4	U	1766–7	U	1788–9
V	1722–3	V	1744–5	V	1767–8	V	1789–0
W	1723–4*	W	1745–6	W	1768–9	W	1790–1

* A mug with a Newcastle stamp of 1723-4 (W) in Messrs. Garrard's possession.

† The Assay Master of Newcastle-upon-Tyne, in his evidence before the Committee of the House of Commons, before alluded to, says expressly: "*The letter for the present official year (1772-3) is D.*"

NEWCASTLE-UPON-TYNE ASSAY OFFICE LETTERS.

CYCLE 5.		CYCLE 6.		CYCLE 7.		CYCLE 8.	
A	MAY 1791–2	a	MAY 1815–6	A	MAY 1839–0	a	MAY 1864–5
B	1792–3	b	1816–7	B	1840–1	b	1865–6
C	1793–4	c	1817–8	C	1841–2	c	1866–7
D	1794–5	d	1818–9	D	1842–3	d	1867–8
E	1795–6	e	1819–0	E	1843–4	e	1868–9
F	1796–7	f	1820–1	F	1844–5	f	1869–0
G	1797–8	g	1821–2	G	1845–6	g	1870–1
H	1798–9	h	1822–3	H	1846–7	h	1871–2
I	1799–0	i	1823–4	I	1847–8	i	1872–3
K	1800–1	k	1824–5	J	1848–9	k	1873–4
L	1801–2	l	1825–6	K	1849–0	l	1874–5
				L	1850–1		

M	1802–3	m	1826–7	M	1851–2	m	1875–6
N	1803–4	n	1827–8	N	1852–3	n	1876–7
O	1804–5	o	1828–9	O	1853–4	o	1877–8
P	1805–6	p	1829–0	P	1854–5	p	1878–9
Q	1806–7	q	1830–1	Q	1855–6	q	1879–0
R	1807–8	r	1831–2	R	1856–7	r	1880–1
S	1808–9	s	1832–3	S	1857–8	s	1881–2
T	1809–0	t	1833–4	T	1858–9	t	1882–3
U	1810–1	v	1834–5	U	1859–0	u	1883–4
W	1811–2	w	1835–6	W	1860–1		
X	1812–3	x	1836–7	X	1861–2		
Y	1813–4	y	1837–8	Y	1862–3		
Z	1814–5	z	1838–9	Z	1863–4		

NOTE.—The usual stamps found upon plate assayed at Newcastle are:—1. The Lion passant. 2. The Leopard's Head crowned. 3. The Town Mark of Three Castles. 4. The Letter or Date Mark; and 5. The Maker's Initials. After 1784 the Duty Mark of the Sovereign's Head is added.

EXAMPLES

A porringer with two handles, fluted base and gadroon border at top. Date about 1680.—*The Earl of Breadalbane.*

Large gravy ladle Date 1725.—*H. A. Attenborough, Esq.*

Ditto. 1740.—*Messrs. Hancock.*

Small beaker. Date 1740.—*The Marquis of Exeter.*

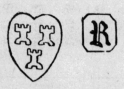

Do. 1746. *Messrs. Hancock.*

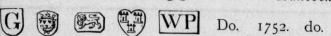

Do. 1752. do.

Do. 1764. do.

Do. 1765. do.

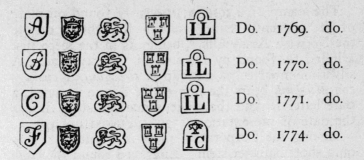

					Do.	1769.	do.
					Do.	1770.	do.
					Do.	1771.	do.
					Do.	1774.	do.

NORWICH.

In Norwich, plate was assayed and marked at an early period, and some specimens are existing among the Corporation plate of the date 1567. An annual letter seems to have been used, for we find on a gilt cylindrical salt and cover, elaborately chased with strap-work and elegant borders, this inscription:— "The Gyfte of Peter Reade, Esquiar, to the Corporation." The plate-marks are—1. The Arms of Norwich, viz., a castle surmounted with a tower, in base a lion *passant gardant;* 2. A Roman capital D.; and 3, Cross-mound (or orb and cross) within a lozenge. It was therefore made and stamped at Norwich before 1568, for Peter Reade died in that year.

Among the records of the Corporation of Norwich we see that in 1624 the mark of a castle and lion was delivered by the Mayor and Corporation to the Wardens and Searcher of the trade of goldsmiths; and on July 1, 1702, Mr. Robert Harstonge was sworn assayer of gold and silver to the Company, although we have never met with any plate with marks of Norwich after that date.

A cocoa-nut cup, mounted in silver, bears the city arms of castle and lion and a rose crowned, with the date mark, a Roman capital S.—*Messrs. Hunt & Roskell.*

The stamp of a rose is frequently found on plate of the sixteenth century, and is thought to denote the Norwich Assay Office, being, as in the piece just alluded to, found by the side of the city arms. A silver-mounted cocoa-nut cup *in the South Kensington Museum* bears the impress of a rose crowned, a date letter R, and the maker's mark, a star. It has the date of presentation, 1576, engraved upon it.

Among the Corporation plate is a gilt tazza cup on a short baluster stem. Engraved round the edge in cusped letters is the following inscription:— "THE MOST HERE OF IS DVNE BY PETER PETERSON." He was an eminent goldsmith at Norwich in the reign of Queen Elizabeth. In the bottom of the bowl are engraved, within a circle, the arms of the city of Norwich, viz., gu. a castle surmounted with a tower ar., in base a lion passant gardant or. Two plate-marks have existed on the edge of the bowl. One of these seems to bear the arms of the city in an escutcheon, which was used to distinguish the plate made and assayed at Norwich, and the other a cross-mound. English work, the latter half of the sixteenth century. There are two other cups of similar character belonging to the Corporation, on one of which are the following assay marks, the lion, leopard's face, a covered cup, and letter. All three were probably the gift of John Blenerhasset whose arms are engraved within one of them. He was steward of the city in 1563, and one of the burgesses in Parliament, 13 Eliz.—*Proceedings Arch. Inst.* 1847.

A silver mace-head of the Company of St. George, in form of a capital of a column, enriched with acanthus leaves, and surmounted by a statuette of St. George and the Dragon. Round the collar has been engraved, but now partly obliterated by the insertion of four sockets, the following inscription:—

"*Ex Dono Honorabil: Fraternitatis Sti. Georgij in Norwico*
An° Do^m 1705."

On the top is engraved the shield of St. George and the following :—

"DIE III. MAEII, MDCCLXXXVI. BENI ET FELICITER MVNICIPIO NORVICENSI OMNIA VT EVENIANT PRECATVR ROBERTVS PARTRIDGE PRAETER."

The plate-mark, a court-hand *b* in an escutcheon on the mace-head, is of the year 1697. The initial H occurs on one of the marks, the remainder of which is illegible. Height 12¾in.—*Ibid.*

The Walpole mace, presented in 1733, was assayed and stamped in London.

A finely ornamented repoussé ewer and salver, with Neptune and Amphitrite, "The gift of the Hon. Henry Howard, June 16, 1663," was stamped in London in 1597. A tall gilt tankard, repoussé with strap-work, flowers, and fruit, and engraved with the arms of Norwich, was stamped in London in 1618.

The rose crowned is the standard mark; the castle and lion that of the town; the cross-mound and star being the mark of the famous Peter Peterson. All the silver bearing this symbol having been made by him.

EXAMPLES.

A chalice dated 1567, stamped with the letter C, and a cross-mound within a lozenge.—*North Creake Church, Norfolk.*

A piece of plate, date about 1567.— *Messrs. Hancock.* Communion cup of the same date.— *Messrs. Hancock.*

Silver gilt salt. Date 1568. — *The Corporation of Norwich.*

  R

Mount of a cocoa-nut cup, with the date of presentation 1576.—*South Kensington Museum.*

A cocoa-nut cup, stamped with a rose, and the letter S, date about 1580.—*Messrs. Hunt & Roskell.*

Seal-top spoon. Date *circâ* 1637.—*J. H. Walter, Esq.*

Split head spoon. Date *circâ* 1662.— *J. H. Walter, Esq.*

   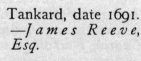

Tankard, date 1691. —*James Reeve, Esq.*

 K

Button top spoon, pounced date 1717, date of make about 1693. — *Messrs. Hancock.*

Beaker, date 1697.—
J. H. Walter, Esq.

SALISBURY.

This city was appointed as an assay town in 1423, but it is not known if plate was ever assayed here; in any case nothing was done in 1700, when several other places were re-appointed as assay towns.

SHEFFIELD.

At Sheffield, silver only is assayed. Mr. B. W. Watson, the Assay Master, has most courteously furnished us with the variable letter for each year from the commencement in 1773, from references to the minute-book wherein are recorded the meetings for the election of new wardens, as well as the letter to be used for the ensuing year. The change takes place on the first Monday in July. The plan adopted at Sheffield differs from all the other offices, for instead of taking the alphabet in regular succession, the special letter for each year is selected apparently at random until 1824, after which the letters follow in their proper order. Through Mr. Watson's kindness, we are enabled to lay before our readers a table of marks, which has been verified by him. The marks used at this office are the same as at London, except that the crown is substituted for the leopard's head, and variation of the date-mark. Sometimes we find the crown and date-letter combined in one stamp, probably on small pieces of plate, but they are generally separate on square punches. When practicable, the four marks are placed in order and struck from one punch, but they are struck separately, when that cannot be done. Occasionally we find the crown and lion on one stamp. The marks are so combined for the convenience of the wardens in marking the goods, but the letter only is used to denote the year in which the article was made.

The date letters are invariably placed in square escutcheons.

The form of the lion and crown now used is:—

SHEFFIELD ASSAY OFFICE LETTERS.

SHEFFIELD ASSAY OFFICE LETTERS.

CYCLE 1.		CYCLE 2.		CYCLE 3.		CYCLE 4.		CYCLE 5.	
1773–4	(crown mark)	1799–0	E (crown)	1824–5	a (crown)	1844–5	A (crown)	1868–9	A
1774–5		1830–1	N	1825–6	b	1845–6	B	1869–0	B
1775–6		1801–2	H	1826–7	c	1846–7	C	1870–1	C
1776–7		1802–3	M	1827–8	d	1847–8	D	1871–2	D
1777–8		1803–4	F	1828–9	e	1848–9	E	1872–3	E
1778–9		1804–5	G	1829–0	f	1849–0	F	1873–4	F
1779–0		1805–6	B	1830–1	g	1850–1	G	1874–5	G
1780–1		1806–7	A	1831–2	h	1851–2	H	1875–6	H
1781–2		1807–8	S	1832–3	k	1852–3	I	1876–7	I
1782–3		1808–9	P	1833–4	l	1853–4	K	1877–8	K
1783–4		1809–0	K	1834–5	m	1854–5	L	1878–9	L
1784–5		1810–1	L	1835–6	p	1855–6	M	1879–0	M
1785–6		1811–2	C	1836–7	q	1856–7	N	1880–1	N
1786–7		1812–3	D	1837–8	r	1857–8	O	1881–2	O

1787–8	𝕬	1813–4	R	1838–9	S	1858–9	P	1882–3	P
1788–9	𝕭	1814–5	W	1839–0	t	1859–0	R	1883–4	Q
1789–0	𝕮	1815–6	O	1840–1	u	1860–1	S	1884–5	R
1790–1	𝕯	1816–7	T	1841–2	v	1861–2	T	1885–6	S
1791–2	𝕰	1817–8	X	1842–3	X	1862–3	U	1886–7	T
1792–3	𝕱	1818–9	I	1843–4	Z	1863–4	V	1887–8	U
1793–4	𝕲	1819–0	V			1864–5	W	1888–9	V
1794–5	𝕳	1820–1	Q			1865–6	X	1889–0	W
1795–6	𝕴	1821–2	Y			1866–7	Y	1890–1	X
1796–7	𝕶	1822–3	Z			1867–8	Z	1891–2	Y
1797–8	𝕷	1823–4	U					1892–3	Z
1798–9	𝕸								

Legend (first group):
1. Lion passant.
2. Crown.
3. Date Letter.
4. Duty, King's Head.
5. Maker's Mark.

Legend (second group):
1. Lion passant.
2. Crown and Date in one stamp.
3. Duty.
4. Maker.

Legend (third group):
1. Lion passant.
2. Crown and Date.
3. Duty.
4. Maker.

Legend (fourth group):
1. Lion passant.
2. Crown and Date.
3. Date. 4. Duty.
5. Maker.
The crown sometimes on a separate stamp.

Legend (fifth group):
1. Lion passant.
2. Crown.
3. Date Letter.
4. Duty.
5. Maker.

For the New Standard, Britannia instead of Lion passant.

SHEFFIELD ASSAY OFFICE LETTERS.

CYCLE 6.

Date	Letter	Date	Letter	Date	Letter
1893–4	𝖆	1896–7	𝖉	1899–0	𝖌
1894–5	𝖇	1897–8	𝖊	1900–1	𝖍
1895–6	𝖈	1898–9	𝖋	1901–2	𝖎

Date	Letter	Date	Letter
1902–3	𝖐	1905–6	𝖓
1903–4	𝖑	1906–7	𝖔
1904–5	𝖒	1907–8	𝖕

1. Lion passant.
2. Crown.
3. Date Letter.
4. Maker.

EXAMPLES.

Candle Stick. Date 1791-2.—*J. H. Wal-ter, Esq.*

Salver. Date 1831-2. —*W. Shoosmith, Esq.*

YORK.

York was one of the most ancient places of assay, and it was mentioned in the Act of 1423. The operations at this place appear to have been discontinued, and it was re-appointed as an assay office in 1700. It does not seem however that much business was ever done here.

It will be seen that in 1772, when a return was made to Parliament, the Assay Office was not in existence; but after that it appears to have recommenced working. In 1848 we find it mentioned as an assay town, but doing very little business.*

The Corporation of the City of York possesses some interesting pieces of plate. A State sword with velvet scabbard, mounted in silver, the arms of the

---

* The last duty paid at the Inland Revenue Office was in July 1869. The officer who formerly acted as assayer for the city of York died many years ago, and no successor has been appointed.

city, emblazoned, the arms of Bowes, &c., of the time of Henry VIII. On the blade is this inscription:— "SYR MARTYN BOWES KNYGHT, BORNE WITHIN THIS CITIE OF YORK AND MAIOR OF THE CITIE OF LONDON 1545. FOR A REMEMBRANCE" (continued on the other side) "GAVE THIS S . . . . TO THE MAIOR AND COMMUNALTIE OF THIS SAID HONORABLE CITIE."

Two tankards, the gift of Thomas Bawtrey in 1673, engraved with the arms of York, were made at York, and stamped with the York mark and the italic capital *P*. The gold cup and other pieces were made elsewhere.

A silver chalice and paten in the Church of Chapel-Allerton, Leeds, has three marks: a half fleur-de-lis and half rose, crowned; an italic *b*, similar to the London date letter of 1619; and maker's initials R.H. On the rim is the date of presentation, 1633.

A stoneware jug has in relief the royal arms of England and the date 1576. It is mounted in silver, and bears three stamps: that of the maker, a half rose and half fleur-de-lis conjoined, and the date letter R; it is in Mr. Addington's collection.

The stamp used at York previous to 1700 was probably that of the half rose and half fleur-de-lis conjoined, which is frequently met with on plate of the sixteenth and seventeenth centuries.

The junction of the lis and rose is probably in allusion to the union of the rival houses of York and Lancaster, by the marriage of Henry the Seventh to the Princess Margaret, daughter of Edward IV., in 1486; the lis being a favourite badge of the Lancastrians as the rose was that of York. As a mint mark we find occasionally the fleur-de-lis on the coins of the Lancastrian kings, in allusion to their French conquests; but upon some of the coins of Henry VII. we find as mint mark the lis and rose conjoined—sometimes half rose and half lis as on the York punch on plate, on others a lis stamped upon a rose, and sometimes a lis issuing from a rose.

The York mark here given, being found on plate of the sixteenth and seventeenth centuries, is frequently much worn and partially obliterated. The half fleur-de-lis is easily distinguished, but the corresponding half is not so easily discerned. In some instances it looks like a demidiated leopard's head crowned; in others like the half of a seeded rose, with portion of the crown over it, for which it is probably intended. There is a great similarity, however, in all the punches we have examined, as if struck from one die, which having been a long time in use may have got damaged. It may be remarked as a curious coincidence, that two diminutive letters can be traced—YO, the two first letters of the word York.

EXAMPLES.

Apostle spoon. Date 1645. *Dallington Church, Northamptonshire.*

Apostle spoon of the seventeenth century. The stamp is a half lis and half rose crowned. Date 1626 —*Rev. T. Staniforth.*

A piece of plate, seventeenth century. —*Messrs. Hancock.*

Ditto.

CW

On a spoon with flat
stem,   leaf-shaped
end and oval bowl,
date about 1680 to
1690. — *Earl   of
Breadalbane.* (This
has also the stamp
of a half lis and
rose, here omitted by
mistake.)

    J

On an oval engraved
teapot. This mark
proves that J was
used as a date letter
previous to 1784, hav-
ing no duty mark.
It may belong to
the year 1736, for J
of 1710 would have
the Britannia mark
of the new standard.
—*Messrs. Hancock.*

Scotland.

EDINBURGH.

THE arms of the city of Edinburgh are: *Ar.* on a rock *ppr.*, a castle triple towered, embattled *sa.*, masoned of the first and topped with three fans *gu.*, windows and portcullis closed of the last. *Crest*, an anchor wreathed about with a cable, both *ppr.* *Supporters:* dexter, a maid richly attired, hair hanging down over her shoulders, *ppr.;* sinister, a doe, also *ppr.* *Motto*, "NISI DOMINUS FRUSTRA."

I.—THE STANDARD.

For Edinburgh—A THISTLE (after 1757); before that, the Assay Master's initials.
For gold of 22 karats, a thistle and 22.
For gold of 18 karats, a thistle and 18.
The present mark is:—

II.—THE HALL-MARK.

For Edinburgh—A CASTLE with three towers, introduced in the fifteenth century (1483); before that the Assay Master's initials.

The three towered castle now used is:—

III.—THE DUTY MARK.

The head of the sovereign, indicating payment of the duty. It was omitted on the debased standards of 15, 12 and 9 karats on gold, although subject to the same duty as the higher standards. Abolished on silver plate.

IV.—THE DATE MARK.

A LETTER OF THE ALPHABET. The custom has been to use the letters alphabetically from A to Z, omitting J, thus making a cycle of twenty-five years (with some exceptions); introduced 1681, and changed on the first Hall day in October every year.

V.—THE MAKER'S MARK.

Formerly some device, with or without the maker's initials: afterwards the initials of his Christian and surname, used from time immemorial, accompanied by the Assay Master's initials only.

1. The standard mark was the deacon's initials from 1457 to 1757, when the thistle was substituted for it.

2. The maker's mark from 1457.

3. The town mark of a castle with three towers from 1483.

4. The date letter from 1681-2.

5. The duty mark of the sovereign's head from 1784, as in England, except on the debased standards of 15, 12, and 9 karats gold, and discontinued on silver plate.

The following table is arranged from the minutes of the Goldsmiths' Company of Edinburgh, where the date-letters appear noted almost every year from 1681, verified by pieces of plate bearing dates. The goldsmiths' year is from Michaelmas to Michaelmas (29th September). The Hall-mark or town mark of a castle was used as early as 1457, and is referred to in that Act (before quoted), and alluded to again in 1483 and 1555.

Previous to 1681, when our table commences, no date-mark appears to have been used. On a piece of plate said to be of the sixteenth century, exhibited at Edinburgh in 1856, in the Museum of the Archæological Institute, we find a castle (the middle tower higher than the two others, as usual), and two other stamps of the letter E. These are, perhaps, the town mark, Assay Master's, and maker's mark. The silver mace belonging to the City of Edinburgh, and known from the town records to have been made by George Robertson in 1617, has three marks, viz., the castle, the cipher G. R., and the letter G. (See p. 181.)

The High Church plate, dated 1643, and the New-battle Church plate, dated 1646, and several others of the same date, have only the town mark, the Assay Master's mark, and that of the maker.

Our thanks are due to the Assay Master, Mr. Alex. Keir, for his kindness in furnishing the present marks.

Our thanks are due here again to the representatives of the late Mr. W. J. Cripps, C.B., for permitting us to include some authorities given by the late Mr. J. H. Sanderson for the Tables of Edinburgh Hall Marks, the property in which had passed to that gentleman.

EDINBURGH ASSAY OFFICE
LETTERS.

EDINBURGH ASSAY OFFICE LETTERS.

CYCLE 1. Black Letter Small.		CYCLE 2. Roman Capitals.		CYCLE 3. Italic Capitals.		CYCLE 4. Old English Capitals.	
OCTOBER		OCTOBER		OCTOBER		OCTOBER	
CHARLES II. 1681–2	a	1705–6	A	1730–1	A	1755–6	A
1682–3	b	1706–7	B	1731–2	B	1756–7	B
1683–4	c	1707–8	C	1732–3	C	1757–8	C*
1684–5	d	1708–9	D	1733–4	D	1758–9	D
JAMES II. 1685–6	e	1709–0	E	1734–5	E	1759–0	E
1686–7	f	1710–1	F	1735–6	F	GEORGE III. 1760–1	F
1687–8	g	1711–2	G	1736–7	G	1761–2	G
WILLIAM & MARY. 1688–9	h	1712–3	H	1737–8	H	1762–3	H
1689–0	i	1713–4	I	1738–9	I	1763–4	I
1690–1	k	GEORGE I. 1714–5	K	1739–0	K	1764–5	K
1691–2	l	1715–6	L	1740–1	L	1765–6	L
1692–3	m	1716–7	M	1741–2	M	1766–7	M
1693–4	n	1717–8	N	1742–3	N	1767–8	N

Date	Letter	Date	Letter	Date	Letter	Date	Letter
1694–5	o	1718–9	O	1743–4	O	1768–9	A
WILLIAM III. 1695–6	p	1719–0	P	1744–5	P	1769–0	B
1696–7	q	1720–1	Q	1745–6	Q	1770–1	C
1697–8	r	1721–2	R	1746–7	R	1771–2	D
1698–9	s	1722–3	S	1747–8	S	1772–3	E
1699–0	t	1723–4	T	1748–9	T	1773–4	F
1700–1	u	1724–5	U	1749–0	U	1774–5	G
1701–2	w	1725–6	V	1750–1	V	1775–6	H
ANNE. 1702–3	x	GEORGE II. 1726–7	W	1751–2	W	1776–7	I
1703–4	y	1727–8	X	1752–3	X	1777–8	K
1704–5	z	1728–9	Y	1753–4	Y	1778–9	L
		1729–0	Z	1754–5	Z	1779–0	M

FOUR STAMPS.
1. The Castle.
2. The Assay Master's Initials.
3. The Maker's Initials.
4. The Date Letter in a pointed shield.

FOUR STAMPS.
1. The Castle.
2. The Assay Mark.
3. The Maker's Initials.
4. The Date Letter in a pointed shield.

FOUR STAMPS.
1. The Castle.
2. The Assay Mark.
3. The Maker's Initials.
4. The Date Letter in a square shield.

FOUR STAMPS.
1. The Castle.
2. The Thistle, in 1757.
3. The Maker's Initials.
4. The Date Letter in a square shield.

From 1700 to 1720 Britannia was added for the New Standard.

* The standard mark of a thistle was used instead of the Assay Master's initials in 1757.

EDINBURGH ASSAY OFFICE LETTERS.

CYCLE 5. Roman Capitals.		CYCLE 6. Roman Small.		CYCLE 7. Old English Capitals.		CYCLE 8. Egyptian Capitals.	
October 1780–1	A	October 1806–7	a	October 1832–3	A	October 1857–8	A
1781–2	B	1807–8	b	1833–4	B	1858–9	B
1782–3	C	1808–9	c	1834–5	C	1859–0	C
1783–4	D	1809–0	d	1835–6	D	1860–1	D
1784–5	E*	1810–1	e	1836–7	E	1861–2	E
1785–6	F	1811–2	f	Victoria. 1837–8	F	1862–3	F
1786–7	G	1812–3	g	1838–9	G	1863–4	G
1787–8	G†	1813–4	h	1839–0	H	1864–5	H
1788–9	H	1814–5	i	1840–1	I	1865–6	I
1789–0	I	1815–6	j	1841–2	K	1866–7	K
1790–1	K	1816–7	k	1842–3	L	1867–8	L
1791–2	L	1817–8	l	1843–4	M	1868–9	M
1792–3	M	1818–9	m	1844–5	N	1869–0	N
1793–4	N	1819–0	n				

Letter	Year	Letter	Year	Letter	Year	Letter	Year
O	1795–6	o	GEORGE IV. 1820–1	𝕺	1845–6	O	1870–1
P	1795–6	p	1821–2	𝕻	1846–7	P	1871–2
Q	1796–7	q	1822–3	𝕼	1847–8	Q	1872–3
R	1797–8	r	1823–4	𝕽	1848–9	R	1873–4
S	1798–9	s	1824–5	𝕾	1849–0	S	1874–5
T	1799–0	t	1825–6	𝕿	1850–1	T	1875–6
U	1800–1	u	1826–7	𝖀	1851–2	U	1876–7
V	1801–2	v	1827–8	𝖁	1852–3	V	1877–8
W	1802–3	w	1828–9	𝖂	1853–4	W	1878–9
X	1803–4	x	1829–0	𝖃	1854–5	X	1879–0
Y	1804–5	y	WILLIAM IV. 1830–1	𝖄	1855–6	Y	1880–1
Z	1805–6	z	1831–2	𝖅	1856–7	Z	1881–2

FIVE STAMPS.	FIVE STAMPS.	FIVE STAMPS.	FIVE STAMPS.
1. The Castle.	1. The Castle.	1. The Castle.	1. The Castle.
2. The Thistle.	2. The Thistle.	2. The Thistle.	2. The Thistle.
3. The Maker's Initials.	3. The Maker's Initials.	3. The Maker's Initials.	3. The Maker's Initials.
4. The Date Letter in a pointed shield.	4. The Date Letter in a square shield.	4. The Date Letter in a shield, concave sides.	4. The Date Letter in an oval.
5. King's Head, 1784.	5. Sovereign's Head.	5. Sovereign's Head.	5. Sovereign's Head.

* In 1784 the Duty Mark of the Sovereign's Head was added.

† The G is repeated, according to the Minutes.

EDINBURGH ASSAY OFFICE LETTERS.

CYCLE 9.

VICTORIA.					EDWARD VII.
1882–3 — a	1887–8 — f	1892–3 — l	1897–8 — q	1901–2 — u	
1883–4 — b	1888–9 — g	1893–4 — m	1898–9 — r	1902–3 — w	
1884–5 — c	1889–0 — h	1894–5 — n	1899–0 — s	1903–4 — r	
1885–6 — d	1890–1 — i	1895–6 — o	1900–1 — t	1904–5 — a	
1886–7 — e	1891–2 — k	1896–7 — p		1905–6 — z	

1. The Castle.
2. The Thistle.
3. The Maker's Mark.
4. The Date Letter.
5. Sovereign's Head until 1890.

EXAMPLES.

George Robertson, maker of the mace of the city in 1617.—*Mr. J. H. Sanderson's Paper, Transactions of the Society of Antiquaries, Scotland*, vol. iv. p. 543, and plate xx.

"On the Dalkeith Church plate there is no date, but it is known from the records to be older than that of Newbattle" (dated 1646).—*Ibid.*

From the plate belonging to Trinity College Church, Edinburgh, bearing date 1663.—*Ibid.* (The castle is omitted by mistake in the cut.)

On a Quaigh, hemispherical bowl with flat projecting handles, on one A C, on the other I M^cL; engraved outside with full-blown roses and lilies. The initials I M^cL are found as a maker on the Glasgow Sugar Castor (p. 97). Date 1713. —*Earl of Breadalbane.*

On a Table Spoon, French pattern, rat's tail. On back of spoon are four marks: (1) maker's unknown; (2) castle; (3) deacon's mark; (4) date-letter U. Date 1749.—*Earl of Breadalbane.*

On a Dessert Spoon, French pattern. The date-letter is the old English $\mathfrak{C}$ of 1757, showing that the thistle was used in this year, as before stated. Maker unknown. Date 1757. —*Earl of Breadalbane.*

Maker's name unknown. Date 1766. —*Earl of Breadalbane.*

Spoon. Date 1837.— *J. P. Stott, Esq.*

GLASGOW.

The arms of the City of Glasgow are: *Ar.* on a mount in base *vert*, an oak tree *ppr.*, the stem at the base thereof surmounted by a salmon on its back also *ppr.*, with a signet ring in its mouth *or;* on the top of the tree a redbreast, and in the sinister fess point an ancient hand-bell, both also *ppr.* *Crest:* the half-length figure of St. Kentigern affrontée vested and mitred, his right hand raised in the act of benediction, and in his left a crosier, all *ppr.* *Supporters:* two salmon *ppr.*, each holding in its mouth a signet ring *ppr.* *Motto*, "LET GLASGOW FLOURISH."

The ancient marks on plate made at Glasgow previous to the Act of 1819, were:—1. The city arms, a tree with a hand-bell on one side and sometimes a letter G on the other, a bird on the top branch, and a fish across the trunk holding a ring in its mouth, enclosed in a very small oval escutcheon. 2. The maker's initials, frequently repeated. 3. A date-letter, but it is at present only possible to assign correct dates for a very few years.

Glasgow was made an assay town by the 59 Geo. III. (May 1819). The district comprised Glasgow and forty miles round, and it was directed that all plate made in the district should be assayed at that office. The peculiar mark of the Glasgow Company is a tree growing out of a mount, with a bell pendant on the sinister branch, a bird on the top branch, and across the trunk of the tree a salmon holding in its mouth a signet ring.

The marks used on the silver plate stamped at Glasgow, since the Act of 1819, are:—

1. *The Standard*, a lion rampant. The present form of which is:

2. *The Hall-Mark*, being the arms of the city, a tree, fish and bell.

3. *The maker's mark*, viz., his initials.

4. *The date-mark*, or variable letter, changed on the 1st July in every year.

5. *The duty mark* of the sovereign's head. Abolished 1890.

For gold of 22 and 18 karats the figures 22 or 18 are added, and for silver of the New Standard Britannia is added.

The Scottish Act of 6 and 9 Wm. IV. (1836-7) in some respects extended to Glasgow, although it is generally regulated by the 59 of Geo. III.; but they have not adopted the marks prescribed by this statute of 1836, and continue those previously in use. The only difference, however, is that the lion rampant takes the place of the thistle.

The lower gold standards of 15, 12, and 9 karats bear the mark of the lion rampant as well as the town mark, being the same as the higher standards, with the difference of quality expressed by numerals.

GLASGOW ASSAY OFFICE
LETTERS.

GLASGOW ASSAY OFFICE LETTERS.

Letter	1st July	Letter	1st July	Letter	1st July
A	1819-0	𝕬	1845-6	A	1871-2
B	1820-1	𝕭	1846-7	B	1872-3
C	1821-2	𝕮	1847-8	C	1873-4
D	1822-3	𝕯	1848-9	D	1874-5
E	1823-4	𝕰	1849-0	E	1875-6
F	1824-5	𝕱	1850-1	F	1876-7
G	1825-6	𝕲	1851-2	G	1877-8
H	1826-7	𝕳	1852-3	H	1878-9
I	1827-8	𝕴	1853-4	I	1879-0
J	1828-9	𝕵	1854-5	J	1880-1
K	1829-0	𝕶	1855-6	K	1881-2
L	1830-1	𝕷	1856-7	L	1882-3
M	1831-2	𝕸	1857-8	M	1883-4
N	1832-3	𝕹	1858-9	N	1884-5

Letter	Year	Letter	Year	Letter	Year
O	1833–4	O	1859–0	O	1885–6
P	1834–5	P	1860–1	P	1886–7
Q	1835–6	Q	1861–2	Q	1887–8
R	1836–7	R	1862–3	R	1888–9
S	1837–8	S	1863–4	S	1889–0
T	1838–9	T	1864–5	T	1890–1
U	1839–0	U	1865–6	U	1891–2
V	1840–1	V	1866–7	V	1892–3
W	1841–2	W	1867–8	W	1893–4
X	1842–3	X	1868–9	X	1894–5
Y	1843–4	Y	1869–0	Y	1895–6
Z	1844–5	Z	1870–1	Z	1896–7

FIVE STAMPS.
1. Lion rampant.
2. Tree, Fish, and Bell.
3. Maker's Initials.
4. Date Letter.
5. Sovereign's Head.

FIVE STAMPS.
1. Lion rampant.
2. Tree, Fish, and Bell.
3. Maker's Initials.
4. Date Letter.
5. Queen's Head.

FIVE STAMPS.
1. Lion rampant.
2. Tree, Fish, and Bell.
3. Maker's Initials.
4. Date Letter.
5. Queen's Head.

GLASGOW ASSAY OFFICE LETTERS.

CYCLE 4.

Year	Letter	Year	Letter	Year	Letter	Year	Letter
1897-8	A	1900-1	D	1903-4	G	1906-7	K
1898-9	B	1901-2	E	1904-5	H	1907-8	L
1899-0	C	1902-3	F	1905-6	J	1908-9	M

1. Lion rampant.
2. Tree, Fish, and Bell.
3. Sovereign's Head until 1890.
4. Date Letter.
5. Maker's Mark.

EXAMPLES OF EARLY MARKS.

GLASGOW. These marks are on the narrow rim of the foot of an elegant silver Tazza, chased in centre with bold leaf scrolls, bordered with engrailed lines. The work is evidently of the time of Charles II., 1670-1680. —*Messrs. Hancock.*

   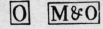

GLASGOW. These four stamps are found on an oval Silver Box, originally made to contain the wax seal appended to a diploma granted by the University. Dated about 1700.—*In the possession of the Earl of Breadalbane.*

GLASGOW. On a Sugar Castor, chased with festoons of roses. This maker's initials are also found engraved on the handle of a Quaigh of Edinburgh, make of 1713.—*The Earl of Breadalbane.*

SCOTTISH PROVINCIAL MARKS.

ABERDEEN.

THE town of Aberdeen bears: *Gu.*, three towers, triple towered, within a double tressure flowered and counter-flowered *arg*. *Supporters*, two leopards *ppr*. *Motto*, "BON ACCORD!"

The Town Assay Office mark adopted at Aberdeen consisted of two or more of the letters in the word, thus the letters A B D, with a mark of contraction above, and later A B D N, as in the following example:

ABERDEEN. On a Table Spoon, handle turned up, and ridges in front of stem, elongated oval bowl, date about 1780. —*Earl of Breadalbane.*

The town arms of three towers, triple towered, sometimes two and one, and sometimes one and two, was also used in the eighteenth century.

BANFF.

A matrix in the office of the Town Clerk of Banff bears an oval-shaped seal of a boar passant, "Insignia Urbis Banfiensis."—*Laing's Seals.*

The mark used in this burgh varied very much, but it generally consisted of the name BANFF, or a contraction thereof.

EXAMPLES.

BANFF. Dessert Spoon, French pattern.—*Earl of Breadalbane.*

BANFF. Dessert Spoon, French pattern, with king's head.—*Earl of Breadalbane.*

BANFF. Table Spoon, French pattern.—*Earl of Breadalbane.*

BANFF. Table Spoon, French pattern, with king's head.—*Earl of Breadalbane.*

DUNDEE (ANGUS)

Arms: Az., a pot of growing lilies *arg. Crest:* A lily *arg. Supporters:* Two dragons *vert*, tails knotted together below shield. *Motto*, "DEI DONUM."

The town mark adopted by the Dundee Assay Office is a pot with two handles containing three lilies, as shown in the following

EXAMPLES.

DUNDEE. On a pair of Sugar Tongs, shell and fiddle pattern, about 1880. —*Earl of Breadalbane.*

DUNDEE. On a Table Spoon, oval bowl, rat's tail, flat stem, leaf-shaped end, date c. 1660. — *Earl of Breadalbane.*

DUNDEE. Tea Spoon, fiddle head, last century.—*E a r l o f Breadalbane.*

ELGIN.

The assay towns of Aberdeen, Inverness, and Banff in the adjoining counties adopted abbreviations of their names, usually the first two or three and the last letters, thus: ABDN, INS, and BA; hence, on the same principle, Elgin used ELN.

The annexed marks are on a Table Spoon, with oval bowl, the end of the handle or stem turned upwards with a ridge down the centre: a form in use from about 1730 to 1760.—*In the Earl of Breadalbane's Collection.*

GREENOCK.

Several marks were used in this burgh. Sometimes a ship in full sail, sometimes an anchor, and sometimes a green oak. The whole of these marks are occasionally found on a single article.

INVERNESS.

There have been goldsmiths in this town since the middle of the seventeenth century. The mark generally used was INS, as a short form of the name of the town. A dromedary or camel, and a cornucopia, were also sometimes employed.

EXAMPLES.

INVERNESS. On a Tea-Spoon, fiddle head, date about 1820, with a cornucopia, the crest of the town of Inverness.—*Earl of Breadalbane.*

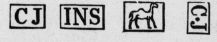

INVERNESS. The camel, one of the supporters of the city arms. On a large annular Scottish Brooch, flat, with engraved vandykes, and a cluster of fine small annulets between each. Maker's mark, and another of the same, larger, as Deacon. Attributed to Chas. Jamison, c. 1810. —*Earl of Breadalbane.*

LEITH.

From the fact of several pieces of plate having been bought here bearing the stamp of an anchor, which indicates its position as a harbour for shipping, we are inclined to attribute this mark to Leith. The circular object with rays, which accompanies it, yet remains to be explained, but in another example here adduced it is placed by the side of the thistle, the standard mark of Scotland. The crest of Edinburgh is an anchor wreathed about with a cable; but in this instance the cable is omitted.

EXAMPLES.

LEITH. Five Tea Spoons, French pattern, 18th century.—*Earl of Breadalbane.*

LEITH. Tea Spoon and Tongs, French pattern, 18th century.—*Earl of Breadalbane.*

LEITH. Caddy Spoon, shell-shaped bowl, fiddle head, with Scottish standard mark and that of a provincial town; no duty letter, but made about 1820, judging from the fashion.—*Earl of Breadalbane.*

LEITH. A Scottish Brooch of conventional form, with circular broad band, plain surface, short pin at back with hinge and clasp; stamped behind with five marks.—*Earl of Breadalbane.*

MONTROSE (ANGUS).

A Burgh Royal, as relative to the name, carries roses. Thus, in the Lyon register of arms—*arg.*, a rose *gules* with helmet, mantling, and wreath suitable thereto.

The town mark, in the seventeenth and eighteenth centuries, was therefore a rose or double rose, in a shield or circle.

PERTH.

The arms of the City of Perth (*alias* St. John's Town) so called since the Reformation, are: An eagle displayed with two heads *or* surmounted on the breast with an escutcheon *gules* charged with the holy lamb *passant regardant*, carrying the banner of St. Andrew within a double tressure, flowered and counter-flowered, *arg.*, with the hackneyed motto, "PRO REGE LEGE ET GREGE."

Goldsmiths have been established in this city from early times.

In the middle of the seventeenth century the town mark was the lamb bearing the banner of St.

Andrew. Somewhat later the double-headed eagle displayed had come into use, and continued to be used until the beginning of the present century.

EXAMPLES.

On a small quaigh, or cup with two handles, date about 1660, with these two marks only. The lamb and flag, emblem of St. John, being the arms of St. John's Town, as Perth was formerly called.—*C. A. North, Esq.*

Split head Spoon. Date *circâ* 1675.— *J. H. Walters, Esq.*

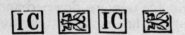

On a set of Table Spoons, French pattern, with rat tail on back of bowl, date about 1760. Some have four marks of spread eagles only, without the shield on the breast, as used recently.—*Earl of Breadalbane.*

On a Dessert Spoon, fiddle head, date *circâ* 1820. The spread eagle part of of the City arms, on its breast a shield with the lamb and flag of St. John; made by Robert Kay, silversmith, at Perth, in 1815.—*Ditto*.

On a set of four Salt-Cellars, gadroon edge on three legs and claws—the seven marks arranged in a circle underneath, with the town mark in the centre, three maker's initials, and three town marks round—date *circâ* 1810.—*Ditto*.

ST. ANDREWS (Fife).

On a matrix of a privy seal in custody of the Town Clerk of St. Andrews is a wild boar passant, secured by a rope to a rugged staff. "Sigillum Secretu Civitatis Sancti Andree Aposti."

Another seal, affixed to a deed dated 1453, bears a full-length figure of a bishop holding a crosier, &c. The counter seal has a figure of St. Andrew extended on his cross. In the lower part of the seal is a wild boar passant, in front of a tree, inscribed around, "Cursus (Apri) Regalis."—*Laing's Ancient Seals*.

STIRLING.

The seal is a lamb couchant on the top of a rock, inscribed with the motto, "OPPIDUM STERLINI."

The ancient seal of the Corporation bears: "A bridge with a crucifix in the centre of it; men armed with bows on one side of the bridge, and men with spears on the other, and the legend, "Hic Armis Bruti, Scoti stant hac cruce tuti."

On the reverse, a fortalice surrounded with trees, inscribed "Continent hoc in se nemus et castrum Strivilense."

"Burke's General Armory" gives the arms of the town, as at present used: *Az.*, on a mount a castle, triple towered, without windows *arg.*, masoned *sa.*, the gate closed *gu.*, surrounded with four oak trees disposed in orle of the second, the interstices of the field being semée of stars of six points of the last, and the motto as above.

The only mark found on silver that can be assigned to this town is a castle triple towered in irregular shield.

STIRLING. On an oblong Tobacco-Box, engraved on the cover with two coats of arms surmounted by a ducal coronet. The town mark is a castle, triple towered, as described above, having beneath the letter S to distinguish it from a similar mark at Edinburgh. The maker's (?) mark, a mermaid and star, and his initials G B.—*Earl of Breadalbane.*

TAIN (ROSS-SHIRE).

TAIN. On a pair of Toddy Ladles, date about 1800.— *Earl of Breadalbane.*

UNCERTAIN SCOTTISH MARKS.

UNKNOWN. These three stamps are on the inside of a silver lid of a shell Snuff-Box. Date about 1800.— *In the possession of the Earl of Breadalbane.*

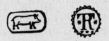

UNKNOWN. On a fiddle-head Toddy Ladle, provincial mark of some town in Scotland. Made *circá* 1810. Representing an otter or badger on a wheat ear (?) and the letters I. & G. H.—*Earl of Breadalbane.*

UNKNOWN. On a seal-top Spoon, of English or Scottish make, of the seventeenth century, the baluster end well finished. The monogram inside the bowl, the animal on the back of the stem. Letters on the bottom, W.E. ᵂ·ˢ· —*Lady Du Cane.* 1621.

UNCERTAIN. (Query Edinburgh). These four marks are on the bottom of a Mug with one scroll handle, broad mouth, repoussé pyriform ornament round the lower part. The small mark is that of the maker, the other two those of the Deacon, probably the same silversmith. Date about 1680.— *Messrs. Mackay & Chisholm.*

Ireland.

DUBLIN.

CHARTER OF INCORPORATION.

THE Goldsmiths' Company of Dublin has the exclusive management of the assaying and marking of wrought gold and silver plate in Ireland.

The harp, and subsequently (A.D. 1638) the harp crowned, was the original Hall or district mark for all Irish manufactured plate assayed in Dublin, and found to be standard, and was used long previous to the charter granted by Charles I., 22nd December, in the year 1638, in the thirteenth year of his reign, to the Corporation of Goldsmiths of Dublin, Ireland. This charter adopted for Ireland the standards then in use in England, viz., 22 karats for gold, and 11 oz. 2 dwts. for silver. "The harp crowned now appointed by his Majesty" has been continued in use ever since, in pursuance of a clause contained in that charter, and also by the Act 23 & 24 Geo. III., c. 23, s. 3 (1784).

The Journals of the Goldsmiths' Company from 1637 until the present time are still in existence, and a complete list of the Masters and Wardens of the Company from that date until 1800 has been printed by Mr. H. F. Berry, M.A., together with the list of Apprentices from 1653 to 1752.

A date mark was used in Dublin from a very early period, as it appears to have been in use previous to the year 1638.

DUBLIN ASSAY OFFICE LETTERS.

The time appointed for the letter to be changed, and the new punches put in commission, was the 29th or 30th, but is now the 24th, May in every year. This date has not been strictly adhered to, the changes having been made at various later periods in some years.

1638 to 1729. 3 *marks:* harp crowned, date letter, and maker's mark.

1806 to 1807. 4 *marks:* harp, date, Hibernia, and maker's initials.

1807 to 1882. 5 *marks:* harp, date, Hibernia, sovereign's head for duty, and the maker's initials.

I.—THE STANDARD (as fixed by the Act 1st June 1784).

For Dublin.—Gold of 22 karats; a harp crowned and the numerals 22.

Gold of 20 karats; a plume of three feathers and 20.

Gold of 18 karats; a unicorn's head and 18.

Silver of 11 oz. 2 dwt.; a harp crowned.

The harp now used is placed in an upright oblong, with the corners cut off :—

No New Standard silver is stamped in Ireland.

THREE LOWER STANDARDS (17 and 18 Vict., 1854).

For Dublin.—*On these the mark of the standard proper (a harp crowned) is omitted, and although subject to the same duty, the mark of the Sovereign's head is also withheld, but Hibernia is used as a Hall-mark.*

    Gold of 15 karats; a stamp of 15.625 (thousandths).

    Gold of 12 karats; a stamp of 12.5 (thousandths).

    Gold of 9 karats; a stamp of 9.375 (thousandths).

For New Geneva.—Gold of 22 karats; a harp crowned *with a bar across the strings* and 22.

    Gold of 20 karats; a plume of *two* feathers and 20.

    Gold of 18 karats; a unicorn's head *with collar on the neck* and 18.

The watch manufactory at New Geneva was discontinued about 1790, having only lasted six years.

II.—THE HALL-MARK.

For Dublin.—A figure of HIBERNIA, used since 1730, on gold or silver of every standard.

The figure of Hibernia is also now placed in a similar outline:—

III.—THE DUTY-MARK.

The SOVEREIGN'S HEAD, first used in 1807 to denote the payment of duty on silver and on the higher standards of gold of 22, 20 and 18 karats; but not on the lower gold of 15, 12 and 9 karats, although paying the same duty. Discontinued on silver in 1890.

IV.—THE MAKER'S MARK.

Formerly some device, with or without the initials of the goldsmith; later the initials of his Christian and surname.

V.—THE DATE-MARK.

From 1638, the year in which the Communion flagon was given by Moses Hill to Trinity College, Dublin, the fact is clearly established, confirmed also by the Charter granted by Charles I. on the 22nd December of 1638, that a Roman letter for that year was adopted commencing with A. No other examples between 1638 and 1679 have come under our notice, but in the latter year we have a chalice with the Old English 𝕭, followed in 1680

by the tankard preserved in the Merchant Taylors'
Company, bearing an Old English 𝕮. Following
the order of the alphabet, plate was doubtless
stamped down to 1686, finishing with 𝕵.

The unsettled state of Ireland during the next six
years will account for the cessation of work at the
Dublin Assay Office. In 1693 the letter 𝕶 (next
in succession) was adopted and continued alphabetically down to 𝕽 in 1700. At this time the Act
of William III., in 1700, reappointing the provincial
offices for adopting the new or Britannia standard,
and making it imperative on all the provincial offices
to discontinue the *old*, may have operated in Dublin,
where the *new* standard was never made, so that a
few years may have elapsed before work was resumed. It appears, from no examples having been
discovered during this period, that in 1710 the Hall
recommenced stamping old standard plate with the
letter 𝕾 next in succession (the top of the shield
being escalloped) down to 𝖅 in 1717, thus com-
pleting the Old English alphabet.

In 1718 a new alphabet was commenced, and as we
have met with two court-hand letters A and C, whilst
Mr. W. J. Cripps ("Old English Plate," edition 1878,
p. 419) gives a letter B in the same hand (although
no authority is quoted in his list of specimens), we
have adopted his suggestion, which is probably correct, viz., that they represented the years 1718, 1719
and 1720.

In 1721 Old English letters were used, and continued with uninterrupted succession (omitting J)
from A to Z, in all twenty-five letters. In 1746
Roman capitals commence, and we have to acknow-
ledge with thanks the late Mr. Cripps' permission to

introduce his arrangement of Roman capitals from 1771 to 1820. It seems unaccountable and contrary to the practice of every other Assay Office to repeat the same character of letter in four successive cycles —the custom has always been to vary the style of alphabet in succession; but at Dublin we have Roman capitals from 1746 to 1845, just a century, the only variations in the Hall-Marks being the introduction of the king's head duty-mark in 1807, and apparently a distinctive form of shield, which, however, was not strictly adhered to throughout each cycle. The arrangement of the tables is still unsatisfactory, and it is to be hoped the promised assistance of the Royal Irish Academy will enable us to clear up the existing discrepancies. Mr. Thomas Ryves Metcalf more than twenty years ago furnished us with extracts from the local Acts of Parliament and extracts from the Minutes of the Goldsmiths' Company recording the Assay Office letters and dates; but he could not do more than give us Roman capitals without any variation of type, hence the present uncertainty, and I am compelled to add, the incompleteness of our Dublin Tables. Mr. S. W. Le Bass, the Assay Master has kindly given us copies of the recent marks.

DUBLIN ASSAY OFFICE LETTERS.

Letter	CYCLE 5. OLD ENGLISH CAPITALS.	Letter	CYCLE 6. ROMAN CAPITALS.	Letter	CYCLE 7. ROMAN CAPITALS.	Letter	CYCLE 8. ROMAN CAPITALS.
A	1721-2	A	1746-7	A	1771-2	A	1796-7
B	1722-3	B	1747-8	B	1772-3	B	1797-8
C	1723-4	C	1748-9	C	1773-4	C	1798-9
D	1724-5	D	1749-0	D	1774-5	D	1799-0
E	1725-6	E	1750-1	E	1775-6	E	1800-1
F	1726-7	F	1751-2	F	1776-7	F	1801-2
G	1727-8	G	1752-3	G	1777-8	G	1802-3
H	1728-9	H	1753-4	H	1778-9	H	1803-4
I	1729-0	I	1754-5	I	1779-0	I	1804-5
K	1730-1	K	1755-6	K	1780-1	K	1805-6
L	1731-2	L	1756-7	L	1781-2	L	1806-7
M	1732-3	M	1757-8	M	1782-3	M	1807-8
N	1733-4	N	1758-9	N	1783-4	N	1808-9

Letter	Year	Letter	Year	Letter	Year	Letter	Year
𝔄	1734–5	O	1759–0	O	1784–5	O	1809–0
𝔅	1735–6	P	1760–1	P	1785–6	P	1810–1
𝔆	1736–7	Q	1761–2	Q	1786–7	Q	1811–2
𝔇	1737–8	R	1762–3	R	1787–8	R	1812–3
𝔈	1738–9	S	1763–4	S	1788–9	S	1813–4
𝔉	1739–0	T	1764–5	T	1789–0	T	1814–5
𝔊	1740–1	U	1765–6	U	1790–1	U	1815–6
𝔋	1741–2	V	1766–7	V	1791–2	V	1816–7
𝔎	1742–3	W	1767–8	W	1792–3	W	1817–8
𝔏	1743–4	X	1768–9	X	1793–4	X	1818–9
𝔇	1744–5	Y	1769–0	Y	1794–5	Y	1819–0
𝔷	1745–6	Z	1770–1	Z	1795–6	Z	1820–1

FOUR STAMPS.
1. Harp crowned.
2. Date Letter.
3. Maker's Initials.
4. Hibernia in 1730.

FOUR STAMPS.
1. Harp crowned.
2. Date Letter.
3. Maker's Initials.
4. Hibernia.

FOUR STAMPS.
1. Harp crowned, Plume, or Unicorn.
2. Date Letter.
3. Maker's Initials.
4. Hibernia.

The three Standards of 22, 20 and 18 karats, directed to be used after 1784, are the Harp, Plume, or Unicorn.

FIVE STAMPS.
1. Harp crowned, Plume, or Unicorn.
2. Maker's Mark.
3. Date Letter.
4. Hibernia.
5. The King's Head in 1807.

DUBLIN ASSAY OFFICE LETTERS.

CYCLE 9. ROMAN CAPITALS.		CYCLE 10. ROMAN SMALL.		CYCLE 11. ROMAN CAPITALS.	
1821-2	A	1846-7	a	1871-2	A
1822-3	B	1847-8	b	1872-3	B
1823-4	C	1848-9	c	1873-4	C
1824-5	D	1849-0	d	1874-5	D
1825-6	E	1850-1	e	1875-6	E
1826-7	F	1851-2	f	1876-7	F
1827-8	G	1852-3	g	1877-8	G
1828-9	H	1853-4	h	1878-9	H
1829-0	I	1854-5	i	1879-0	I
1830-1	K	1855-6	k	1880-1	K
1831-2	L	1856-7	l	1881-2	L
1832-3	M	1857-8	m	1882-3	M
1833-4	N	1858-9	n	1883-4	N

Letter	Date	Letter	Date	Letter	Date
O	1834–5	o	1859–0	O	1884–5
P	1835–6	p	1860–1	P	1885–6
Q	1836–7	q	1861–2	Q	1886–7
R	1837–8	r	1862–3	R	1887–8
S	1838–9	s	1863–4	S	1888–9
T	1839–0	t	1864–5	T	1889–0
U	1840–1	u	1865–6	U	1890–1
V	1841–2	v	1866–7	V	1891–2
W	1842–3	w	1867–8	W	1892–3
X	1843–4	x	1868–9	X	1893–4
Y	1844–5	y	1869–0	Y	1894–5
Z	1845–6	z	1870–1	Z	1895–6

FIVE STAMPS.
1. Harp-crowned, Plume, or Unicorn.
2. Maker's Mark.
3. Date Letter.
4. Hibernia.
5. Sovereign's Head.

FIVE STAMPS.
1. Harp-crowned, Plume, or Unicorn.
2. Maker's Mark.
3. Date Letter.
4. Hibernia.
5. Queen's Head.

FIVE STAMPS.
1. Harp-crowned, Plume, or Unicorn.
2. Maker's Mark.
3. Date Letter.
4. Hibernia.
5. Queen's Head.

DUBLIN ASSAY OFFICE LETTERS.

CYCLE 12.
OLD ENGLISH CAPITALS.

Date	Letter	Date	Letter	Date	Letter
VICTORIA. 1896-7	𝕬	1900-1	𝕰	1904-5	𝕴
1897-8	𝕭	EDWARD VI 1901-2	𝕱	1905-6	𝕶
1898-9	𝕮	1902-3	𝕲	1906-7	𝕷
1899-0	𝕯	1903-4	𝕳	1907-8	𝕸

Date	Letter
1908-9	𝕹
1909-0	𝕺
1910-1	𝕻
1911-2	𝕼

1. Harp crowned, Plume, or Unicorn.
2. Maker's Mark.
3. Date Letter.
4. Hibernia.

EXAMPLES.

Two Tankards presented in 1680 to the Guild of St. John. Date 1680-1. —*Merchant Taylors' Company.* And a Box with scroll feet.—*Dublin Exhibition.*

Piece of Plate. Date 1725-6. — *Messrs. Hancock.*

Mace, dated 1728. The top embossed with the royal arms. — *Messrs. Hancock.*

Two-handled Cup. Date 1739-0.— *Messrs. Hancock.*

Silver gilt Sugar Sifter. Date 1785-6.—*J. H. Walter, Esq.*

Spoon. Date 1803-4. — *J. P. Stott, Esq.*

FINIS

DISTRIBUTORS
for the Wordsworth Reference Series

**AUSTRALIA, BRUNEI,
MALAYSIA & SINGAPORE**

Reed Editions
22 Salmon Street
Port Melbourne
Vic 3207
Australia

Tel: (03) 646 6716
Fax: (03) 646 6925

**GERMANY, AUSTRIA
& SWITZERLAND**

Swan Buch-Marketing GmbH
Goldscheuerstraße 16
D-7640 Kehl am Rhein
Germany

GREAT BRITAIN & IRELAND

Wordsworth Editions Ltd
Cumberland House
Crib Street
Ware
Hertfordshire SG12 9ET

INDIA

Om Book Service
1690 First Floor
Nai Sarak, Delhi - 110006
Tel: 3279823/3265303
Fax: 3278091

ITALY

Magis Books
Piazza della Vittoria 1/C
42100 Reggio Emilia

Tel: 0522-452303
Fax: 0522-452845

NEW ZEALAND

Whitcoulls Limited
Private Bag 92098, Auckland

**SOUTH AFRICA, ZIMBABWE
CENTRAL & E AFRICA**

Trade Winds Press (Pty) Ltd
PO Box 20194, Durban North 4016

USA, CANADA & MEXICO

Universal Sales & Marketing
230 Fifth Avenue
Suite 1212
New York, NY 10001 USA

Tel: 212-481-3500
Fax: 212-481-3534